TO MAXIMIZE THE POTENTIAL OF YOUR
&
TO MAINTAIN YOUR HARD EAR

The

Recognition

&

Control

of

Stress

Induced

Weight

Loss

Saboteurs

Other Books Written by Dr. Joel S. Shain

Life Empowerment
Life Without Arthritis
The Energy Drain Syndrome
How to Survive the Poisoning of America
The Whiplash Syndrome: Erasing the Trauma

ELIMINATING
THE
STRESS
OF
WEIGHT LOSS

**THE RECOGNITION AND CONTROL OF
STRESS INDUCED WEIGHT LOSS SABOTEURS**

Dr. Joel S. Shain

The Monterey Wellness Center
335 El Dorado Street, Monterey, California 93940 (408)646-0555

Library of Congress Catalog Card Number: 91-090248

Shain, Joel S.
Eliminating the Stress of Weight Loss -
The Recognition and Control of Stress Induced Weight Loss Saboteurs

1. Weight Loss 2. Stress 3. Diet 4. Self-Help Manual
5. Includes Glossary, Index and Directory
I. Title II. Title:

Shain, Joel S., 1944 -
Copyright © 1992 Joel S. Shain
p. cm.
ISBN 0-9625958-7-x

Although the author and publisher have exhaustively researched all sources to ensure the accuracy and completeness of the information contained in this book, we assume no responsibility for errors, inaccuracies, omissions or any other inconsistency herein. Any slights against people or organizations are unintentional. Readers should consult their own physicians before starting this or any other weight loss program. The Biofeedback Reflex Test will allow the reader to observe a complex neurological mechanism at work. Tests have repeatedly shown that quantities of alcohol and body chemistry altering drugs alter the results obtained. If any of these substances are part of your life, your readings may not be accurate.

Publisher: The Monterey Wellness Center, 335 El Dorado Street, Monterey, California 93940

PREVENT DAILY STRESS FACTORS FROM SABOTAGING WEIGHT LOSS GOALS!

FINALLY & FOREVER

YOU CAN ELIMINATE PERSONAL STRESS FACTORS
THAT HAVE BEEN SABOTAGING WEIGHT LOSS GOALS!

YOU CAN ELIMINATE YOUR:

DIETARY DISTRESS

UNSUSPECTED STRESS FOODS

FOOD REJECTIVITY PROBLEMS

UNSUSPECTED FAT PRODUCING STRESS FOODS

TOXIC WASTE BACK-UP PROBLEMS CAUSED BY DIETING

UNCONTROLLABLE CONDITIONED NEGATIVE EATING REACTIONS

FOOD SABOTEURS THAT TRIGGER YOUR FOOD INSTINCT SIGNAL

FOOD ADDICTIONS THAT ARE TRIGGERED BY ANXIETY, STRESS OR TENSION

WEIGHT LOSS SABOTEURS CAUSED BY HEREDITY, LIFESTYLE, STRESS AND ILLNESS

FACIAL LINES AND WRINKLES CAUSED BY DIETING, ANXIETY, STRESS OR TENSION

HERE'S WHAT REVIEWERS HAVE TO SAY ABOUT

ELIMINATING
THE STRESS OF WEIGHT LOSS

THE RECOGNITION AND CONTROL OF
STRESS INDUCED WEIGHT LOSS SABOTEURS

"Dr. Shain brings together that host of hard-to-capture, subtle culprits of life that victimize us without our ever understanding how captured we have become. It is a remarkable job of bringing to light those unmanageable life forces we may have succumbed to, understanding them and dealing with them in an intuitive, rational, sane and sensitive way."
Dr. B.B. Brown, originator of the science of Biofeedback, author of *Stress and the Art of Biofeedback*; and *Between Health and Illness*

"Unparalleled .... Dr. Shain, with steady courage and intellectual discrimination, has superbly explained the weight loss challenge."
Dr. Scott Walker, founder and developer of the *Neuro-Emotional Complex*

"This is a revolutionary book about how we undermine ourselves and what we can do about it."
Marilyn Ross, author of *The Complete Guide to Self Publishing*

"An impressive and thoughtful analysis of how our natural stress defense mechanisms sabotage weight loss goals."
Dr. Khelly Webb

"Dr. Shain presents fascinating new concepts that represent a radical, unprecedented look at our natural stress defense mechanisms."
Mr. Ed Dunnly, Product Development Manager, The Chattecx Corp.

"Each chapter is vital for successful dieting."
Yvonne Nienstadt, Director of the Cal-a-Vie Resort/Spa

"The first real improvement in dieting concepts since calorie counting." Jone Chappell, C.E.O., Life-Time® Fat-Free Cheeses

"A unique and innovative approach to weight loss. The creative genius of Dr. Shain will take you to new heights of awareness regarding weight loss issues."
Mark Victor Hansen, author of *Dare to Win;* and *Future Diary*

"A radical, fascinating whole-person approach to weight control. Its benefits will extend into all areas of one's life."
Michael Blate, author of *The Tao of Health*

"An invaluable work, needed by anyone who wishes to make their way through the maze of diets to reach the goal of true health and fitness."
Dr. Stephen T. Chang, author of *The Complete System of Self-Healing*

"Dr. Shain's book unlocks those weight loss secrets that will create many positive changes in the dieter's life."
Dr. John F. Thie, author of *Touch For Health*

"A rare book that provides immediate insights into the dieter's self-empowering self-control resources. Dr. Shain's ability to present this information in an entertaining and exciting way is refreshing."
Dr. William Renas, National Center For The Transitional Healing Arts

"A major breakthrough that takes the mystery out of weight loss." Michael Reed Gach, author of *Acupuncture's Potent Points*

"Your program to recognize and control stress induced saboteurs is a valuable addition to every dieter's library."
Justene Moore, Director - The Touch For Health Foundation 1976-77

"The information in Dr. Shain's book has helped me give my world-class athletes the winning edge."
Dr. Michael Greenberg - 1984, 1988 & 1992 Olympic Team Physician

DEDICATION

This book is dedicated to my closest and dearest friend who is also the love of my life - Katie. Without her help and encouragement, this project would not have been possible. Her love, her warmth and her devotion have all combined to give me the faith, hope and belief that I could complete this project. I extend my sincere gratitude to her for her creative input, suggestions, ideas and unconditional loving. I cherish the memories and look forward with great joy to our years of marital bliss.

ACKNOWLEDGMENTS

I would like to express my appreciation for all of the marvelous people of the Touch For Health Foundation and the doctors of the International College of Applied Kinesiology for laying the groundwork for much of the information presented in this book. Dr. John F. Thie, Dr. George F. Goodheart, Dr. David Walther, Dr. Khelly Webb, Dr. Willie May, Dr. David Denton, Dr. John Diamond, Dr. J.W.S. Hawkins, Dr. George Eversaul, Dr. Thomas Wing, Dr. Mary Marks, Gordon Stokes, Justine Moore and of course Grace Baldridge.

Also to be acknowledged are the doctors and volunteers of The Monterey Wellness Center who created an atmosphere of loving, healing and education. Dr. Paul Andrews, Dr. Mike Greenberg, Annie Rainbow, Dr. Mike Neuman, Dr. Steve Gross, Dr. William Renas, Glenda Anderson, Linda Marin, Duane Judge, Rev. Dori D'Angelo, Lois Ferst, Jonie Chappell, Elaine Jones, Martha Aldridge, Ruth Jones, Julie Temple, Erin Reilly, Holly Urich, Maricela Meers, Ann Cassara, Barbara Franklin, Walt McDonnell, Bill Graham, Lois Sawyer, Amber K. Magsayo, Mary K. Benson, Bob DeLong and the kind folks at Copy King and Kinko's.

Also to be acknowledged are the thousands of Monterey Peninsula residents who attended our self-care classes and worked side by side with our doctors to develop the self-help classes for which the Center became so famous.

AND SPECIAL THANKS TO

Bruce Wright and Maureen Parrott for their creation of this book's design that has allowed me to express myself in a most unique and helpful manner; Maureen Parrott for her creative graphics that bring life to my words; Irene Montagna for her editing skills in making the thoughts fluid; Karen Callaghan for her patience in typing the manuscripts many times; Susan Elizabeth for her final proofing skills; Bunny Hartman for the cover design and again to Bruce Wright who taught us all how to use our desktop publishing programs.

TABLE OF CONTENTS

"Expecting to
Lose Weight
Through Restriction
of Diet Alone,
is
Like Treating
a
Single Symptom
of a
Complex Disease."

Dr. Joel S. Shain
Eliminating the Stress of Weight Loss

INTRODUCTION

Why I Wrote This Book

The Recovery Process

The Stress of Weight Loss

What You Will Find In This Book

How You Will Benefit From This Book

How To Use This Book

All Things Considered

INTRODUCTION

Why I Wrote This Book

As far back as I can remember, I have been fascinated by the human body and the study of biology, anatomy, and the physical sciences. In 1967, I received my degree in Biology from Temple University in Philadelphia, Pennsylvania. As a graduate student, one of my tasks was to take complicated research material and organize it in a way that the research professors could present it to the undergraduate classes. Very quickly, other professors in the science department discovered my ability to simplify their complicated research material. In 1968, my first professional assignment outside the University was as Science Coordinator for a Government Research Project. Because of the success of the Sputnik space satellite, it was decided that American school children had to catch up to the Russians in the study of sciences. I was asked to take the new advances in science and rewrite the science curriculums for grades kindergarten through eighth. Again, my job was to take complicated information and present it in a way that elementary school children could accept and understand.

It was during this time that I had to "watch" my mother die from cancer. I use the word "watch", because although I could describe in detail every aspect of my mother's disease (and the rationale for the treatment she was receiving), there was nothing I, or any of the many loving friends and relatives could do to help ease her pain and suffering. We did not know how, nor was anyone else available to tell us how we could use our hands, our hearts, or our minds to help her during that difficult time. We were living the lie of denial. "Hang in there until that "magic drug" comes along; this isn't really happening; this will pass and you will be fine".

But, the nagging questions remained, "What could the family do to ease her pain and suffering? How could we help to soothe her body, mind and spirit? What could we do to make the time still available more valuable and meaningful? What could we do to relieve her anxiety, stress, tension and depression?"

Before I could learn the answers to these questions, my mother passed away.

With her passing, I decided that I still owed it to her memory to continue this quest in order to help others who did not have the time to research these matters. I embarked on a great adventure into the realm of healing. During this period, I studied the recorded information detailing 5,000 years of health and healing methods. My scientifically trained mind was able to eliminate many of the methods that would not have any success on healing based on what we presently know about human body functions. But, I also was able to find self-care treatments that were significantly better than those available in our present society. As an educator, I felt that it was my responsibility to organize, report and teach these observations.

My first chance to utilize all of this information came in 1974 when I was hired by Dr. John F. Thie to help organize and teach his marvelous program -- Touch for Health. It is estimated that since that time, two million people have studied this program on acupressure, touch and massage as it applied to health and healing.

In 1976, I was asked to assist the Sacro Occipital Teachers Organization. My task was to first review the already successful Sacro Occipital Technique program that was being taught to doctors throughout the country. Then, I was to increase the effectiveness of their classes by reorganizing the material into a more easily understood format.

In 1978, I was appointed Director of the National Center for the Transitional Healing Arts. This organization presented information and postgraduate symposiums on alternative healing methods that were not taught to doctors in medical school.

With this information, physicians were able to provide alternative healing methods to interested patients. Also in 1978, I assisted the Dental Kinesiology Foundation in teaching dentists about alternative healing practices.

Finally in 1978, I decided to fulfill my initial mission - to present alternative health/healing information on a personal, one-on-one basis in a clinical practice. It was then that I opened the doors of The Monterey Wellness Center in Monterey, California. This healing facility offered the creative talents of medical, chiropractic, osteopathic, and naturopathic physicians. There were therapists, counselors and even, on occasion, spiritual healers. My goal was to help the chronically ill learn what they and their family members could do (outside of what was already available medically) to help ease and relieve their suffering. My second goal was to provide the tools for self-aided recovery, self-responsibility and self-respect.

I knew this decision was correct when I met our first patient, Gail B. from Pebble Beach, California, a woman about my mother's age. She walked in the door and said, "I have cancer and I need someone to tell me what my husband and I can do to become involved in my recovery process."

The Recovery Process

The Wellness Center was created to help people with chronic degenerative diseases and other life-threatening problems. We provided these people with information that was not provided by their medical physicians. Their doctors were preoccupied with monitoring the progress of the disease and its responses to various medications or surgical procedures. The doctors at the Wellness Center, on the other hand, were teaching patients and their families what they could do (in addition to what their physicians were already doing) to help in their recovery process. They were taught whatever was necessary to soothe and heal their body, mind and spirit.

Patients came from around the country to aid in their recovery from surgery; they came to the Center to build their strength prior to surgery; they came to the Center to see if they could enhance their health so that surgery might not be necessary; they came to the Center to discover if there was anything in their lifestyle that was adding to their present problem; and they came to the Center to explore the role played by their many loving friends and relatives in helping them recover. But mainly, they came to the Center to learn self-care treatments that they could do to help to speed their recovery.

Diet was one of the first areas explored with each individual: Were there foods in their diet that were aggravating their problem?; Were there foods in their diet that were causing their problem?; Were there foods that could be added to their diet to help alleviate and heal the individual's problem? Early in my research of self-help healing methods, I discovered that there were many good and effective diet programs available for each of the illnesses treated at the Center. But, before I could use them effectively, the following questions had to be answered: Which of the many diet programs available for a specific problem would be best for this specific individual?; What made one of the many diet programs available for a specific problem more beneficial than others?; What factors made the same diet disastrous for one individual but great for another?; Why in developing a so-called "stress-free diet", did we find that foods good for one person were not necessarily good for another?

Eventually, what I discovered was, just as we all look different physically, we are physiologically different as well. It became clear to me that four determining factors played a key role in how an individual would respond to a specific healing dietary regimen. The four determining factors were an individual's heredity, lifestyle, daily stress factors and the severity of the illness. Because of these life factors, certain foods that might be one-hundred percent beneficial to one individual, could represent an Unsuspected Stress Food for another. Because of an individual's life factors, a food that was normally health-giving, could require more energy to digest than it would ever give up in restorative value to the ill body. Then I made a great discovery. I could use biofeedback to teach patients how to recognize those foods, specific only to them, that should be eliminated from their individual dietary program. Thus, a person with a specific problem could be instructed on the optimum diet for that problem and then be taught a simple biofeedback technique enabling him to recognize which foods in that program to avoid. If eaten, these foods would sabotage the health goals of the diet. People using this biofeedback technique recovered much faster than test groups that did not use it. With this biofeedback technique I was able to dramatically increase the benefits of a patient's dietary regimen. The dietary programs became more successful because I eliminated the Unsuspected Stress Foods specific to each individual that were sabotaging his health goals.

Another health issue that had to be overcome was the problem of toxic waste products generated from the illness. Every illness produces toxic waste products that must be eliminated. The body is built to handle this by utilizing the kidneys, the liver, and its other filtering/elimination systems. But, when individuals maintain a disease state for more than a few weeks, the elimination systems can become clogged and overwhelmed. They operate at a slightly inferior level than a normal healthy system. Cleansing the body of the toxic waste products of disease proved to be another significant health enhancement procedure. Although each disease process had its own specific type of toxic waste, the methods to enhance elimination, filtration and cleansing were similar. I found that, in many instances, recovery from disease was delayed because the elimination processes of the body were so overwhelmed. The debris of illness stayed in the body causing secondary problems in other areas. These secondary problems (complications), left untreated, made the recovery process much more difficult. Teaching patients safe and simple methods to help in this cleansing process resulted in a faster recovery.

The third area that had to be confronted with chronic disease was the effects of personal addictions on the patient's present state of health. The most obvious addictions that brought people into the clinic were smoking addictions (lung, throat and other metabolic disorders), alcohol addictions (kidney and liver disorders), food addictions (diabetes and arthritis) and activity addictions (chronic fatigue and stress). But, what I found, was that there were many other "unrecognized" addictions that were, if not directly causing the illness; certainly allowing the illness to stay within the body. I discovered that many people were addicted to those foods, lifestyles, substances, environments, etc., that were causing their problems. Thus, anti-addiction procedures were needed for irritants such as cigarettes, alcohol, foods, and other addictive stimulants/activities. By eradicating the addictions that were causing physical distress to the ill body, many serious illnesses were eliminated.

The fourth area that was vital in the recovery process was the elimination of negative subconscious behavioral patterns resulting from conditioning. Many ill people were sabotaging their recovery processes by reverting to behavior developed in early childhood based on fears, anxieties and the limited knowledge of childhood. They were plagued by subconscious thoughts of inadequacies; lack of self-worth; lack of self-esteem and lack of recognition of their true light of being. Almost immediately, I realized that it was not necessary to dig up and discuss these old negative pictures. What was needed, I discovered, was to erase and to replace those negative pictures with enlightening and empowering visual images and desires. It was simpler to create new positive life enhancing images that would allow the person to move on. With these new subconscious images, individuals were freed from old negative pictures and concepts. This freedom allowed their automatic self-healing mechanisms a greater opportunity to contribute to the recovery processes.

The Stress of Weight Loss

You might be asking yourself what does all this have to do with weight loss or maintaining weight at a consistent level? My observations showed that the health enhancement programs taught at The Monterey Wellness Center had benefits beyond enhancing the health and wellness of the individuals. I noticed that many of my patients were able to reach and to maintain long sought-after weight loss goals. Many were overweight; individuals who had tried multiple diets through the years of their illness and failed. Patients regularly told me that aside from feeling better and being healthier that they had lost the weight that they were trying to lose; or were finally able to maintain weight at consistent levels. With this kind of feedback, I decided to explore the weight loss question. My research revealed immediately that expecting to lose weight, and even more importantly, expecting to maintain hard-earned weight loss goals through diet alone, is like treating a single symptom of a complex disease. A weight loss program can only achieve long-lasting results when it takes into account all of the factors that contribute to making the body what it is. Many individuals on the weight loss roller coaster are, in reality, suffering from a disease. It is the kind of disease that will wreck an individual's life. Until this is understood, individuals will continually try one of the many excellent weight loss programs available; lose weight; put the weight back on; and then go through periods of guilt, remorse, self-doubt, and low self-esteem. Out of frustration, after their weight returns, they will then try another diet -- which will work -- they will lose weight. Eventually though, the same life factors that sabotaged health goals will sabotage that individual's weight loss goals again. People become locked in a cyclic pattern of failure as they fluctuate through periods of denial and then failure/guilt. They never achieve independence from the weight "problem". When I started to classify the weight loss syndrome as a disease, individuals who were interested, were put on the same program as patients with diseases such as arthritis, diabetes, anxiety/stress problems, liver and gallbladder problems or cancer -- The Health Enhancement Program of The Monterey Wellness Center.

The first step for an individual wanting to deal with a weight problem is to learn to recognize their specific Unsuspected Stress Foods. These are foods that are not being digested properly because of heredity, lifestyle, stress or illness. Teaching patients simple, safe biofeedback techniques to recognize problem foods prior to eating them, greatly enhanced the benefits of a weight loss regimen.

Secondly, these patients learned that all diets produce excessive toxic waste by-products created by the destruction of fat cells during dieting. This debris must be excreted by the body's purification and elimination systems. During this process these systems can become overwhelmed. How quickly it happens depends on heredity, lifestyle, daily stress and illness. When it does happen, a situation known as Toxic Waste Back-Up occurs. It is this problem which causes individuals to become aware of a toxic-like mild illness condition. Left untreated, this factor causes many people to abandon a weight loss program because they are "just not feeling quite right" as a result of the program.

Thirdly, dieters must learn how to eliminate their addictions. Many dieters are addicted to those foods which caused their weight problem in the first place. Many are plagued with conscious and subconscious cravings for these foods that eventually cause them to sabotage their weight loss goals. These addictions surface to cause problems during the diet program whether dieters are trying to lose weight or to maintain hard-earned weight loss goals. No individual will ever be free from the cycle of weight gain/weight loss until they erase their personal food addictions.

And finally, dieters must realize that when they were young and growing up, they were very susceptible to peer group pressures. They were susceptible to snide comments from people who were trying to keep them in place. They were susceptible to negative conditioning which influenced their concepts of self-worth and self-image. Many individuals consistently need weight loss regimens to help negate the effects of poor self-image and negative conditioning developed during childhood. Many weight loss programs eventually fail because individuals believe that they are not worthy of having the kind of body they want. Subconsciously, they believe that they are not worthy of having fun, of being beautiful or of the admiration of others. Because of these subconscious beliefs, dieters who achieve their weight loss goals often fall prey to long-stored negative conditioning and behavioral patterns that subconsciously sabotage the most sincere weight loss desires.

The book you are reading, *Eliminating the Stress of Weight Loss* is adopted directly from The Health Enhancement Program of The Monterey Wellness Center. Originally, this program was designed to enhance the health, the life, and the well-being of patients at the Wellness Center. With slight modification, I adapted the program to healthy people wishing to get off the weight loss roller coaster. I believe that the weight loss problem in this country needs to be treated as a disease, but not through the route that is looking for that one "magic formula"-- a "magic formula" that takes the shape of pre-processed weighed food, powdered diet drinks, weight loss pills or injections of embryonic fluid, B12 or whatever "magic bullet" being touted at the moment. After sharing my findings about weight loss with you, I ask my readers the same question that I asked my patients: "Do you think you can control this complicated disease by treating a single symptom?"

This book shows you how to meet your goals and aspirations as well as how to maximize the effectiveness of your present weight loss program. It shows you how to maintain your ideal weight for the rest of your life through pointing out those saboteurs that are playing havoc, not only with your weight problems, but with your health and your life. By knowing your saboteurs, you can avoid the guilt-failure cycle of the Weight Loss Syndrome. With this information, not only will you understand why you always regain weight lost in a weight loss program but you will see how stress induced weight loss saboteurs have prevented you from enjoying your life.

What You Will Find In This Book

This book shows the dieter how to recognize and to eliminate eight of the most significant weight loss saboteurs that are caused by the body's natural stress defense mechanisms. They are: 1) Unsuspected Stress Foods; 2) Food Rejectivity; 3) Food Addictions; 4) Toxic Waste Back-Up; 5) Poor Self-Confidence; 6) Unconscious, Negative Reaction Behavior Reactions; 7) Negative Energy Drains; and 8) Prior Conditioning. It is rare to find an individual affected by all of these weight loss saboteurs. But, I am sure that many persons reading this book will find at least one or two affecting their life or their weight loss goals. Some are being affected by three or four of these saboteurs and, 25 percent of you reading this book will find more than four saboteurs affecting you in one way or another. With this in mind I wrote the following short summaries for each chapter to help you recognize the benefits you can derive from this book.

Chapter One presents the first descriptions of the six solutions needed to eliminate those stress induced weight loss saboteurs triggered by the body's natural stress defense mechanisms. These mechanisms, as detailed in Chapters Two, Six, Eight and Nine can be controlled once the dieter is made aware of them. The recognition and control of these natural defense mechanisms will free the dieter from the stress of weight loss.

Chapter Two discusses the stress mechanisms at play in this weight loss drama. In this chapter, you will see how heredity, lifestyle, stress and illness can create negative reaction pathways within your nervous systems. Once created, these reaction pathways operate on automatic pilot -- without conscious control. These internal mechanisms can then function improperly because of stress factors. The mechanism by which you become addicted to a specific food is presented here. The understanding of how your life's daily stressors affect dieting will start the process to free you from the weight loss roller coaster. Aside from sabotaging your weight loss goals, these reaction pathways (set up within your nervous system) can also sabotage other aspects of your daily life making your daily life just a little more difficult than it really needs to be.

Chapter Three introduces the Unsuspected Stress Foods that trigger your Food Instinct Signal. The Food Instinct Signal is a function of your automatic nervous system. Some people call this their "intuition". You will learn about the Food Rejectivity Syndrome. This syndrome results when your body wants to reject a certain food. The discovery that your Food Instinct Signal is triggered because of the Food Rejectivity Syndrome, represents a major advancement in the understanding of the weight loss roller coaster. It is this information that will allow you to recognize your personal Unsuspected Fat Producing Stress Foods.

Chapter Four teaches you a fascinating biofeedback reflex that will allow you to recognize your Food Instinct Signal. Using this biofeedback technique to monitor your Food Instinct reflex, you will be able to observe your personal responses to various foods. This chapter presents the details of the Biofeedback Reflex Test (B.R.T.) used for the process.

By the time you get to **Chapter Five,** you will know how to use the Biofeedback Reflex Test to monitor your Food Instinct Signal. This chapter will then take you through sample menus of breakfast, lunch and dinner to show you how to use this technique to discover your Unsuspected Fat Producing Stress Foods.

Chapter Six explains a sixty-second anti-addiction procedure the dieter can use to erase the "cravings" for foods created through the stress/addiction process. This technique represents a compilation of six of the most effective and fascinating self-care activities for soothing and restoring those stress mechanisms responsible for addictions.

Chapter Seven continues the food addiction discussion and teaches you how to do the entire anti-addiction procedure. This sixty-second technique helps subdue any of your food "cravings" triggered by stress, tension or anxiety. It is used during the day when you become aware of "cravings" for foods that you should not be eating. After 30 days of this kind of conditioning, the "craving" sensations of many food addictions can be totally erased.

In **Chapter Eight** you learn how to eliminate the debris and waste products of fat cell destruction. We "lose" weight by "destroying" unwanted fat cells. It is the debris of this destruction that can cause Toxic Waste Back-Up. Chapter Eight presents a 14-day cleansing program that helps the elimination and purification systems of your body to operate more effectively, as well as an acupressure treatment to aid digestion and elimination. As an added bonus, this chapter includes methods that improve facial complexion by removing even more of the toxic wastes from your system.

In **Chapter Nine** you learn how to release negative behavioral reaction patterns that have been built into your subconscious memory. Stored deep within your subconscious memory are many old, negative behavioral reaction patterns that are no longer needed in your present day life. Subconsciously activated during stress periods, these stored behavioral reaction pattern memories can be triggered to conscious action. The dieter finds himself overcome by old feelings of poor self-confidence. Also, negative behavior problems that the dieter thought he conquered years earlier re-surface for "no apparent reason". This condition forces him to set low goals. This results in low goals being achieved, which only confirms the dieter's feelings of poor self-confidence. It is this cycle that has to be ended for any dieter to be able to maintain weight loss goals at a consistent level.

Chapter Ten presents the Cosmetic Acu-Point Facial Toning Program of The Monterey Wellness Center. Since most people want a weight loss program to improve their appearance, enhanced facial beauty is a marvelous incentive for maintaining weight loss goals. This chapter demonstrates the use of acupressure to smooth out and tone underlying muscles (not just the top skin layer) of the face. This results in the removal of facial lines and wrinkles caused by stress or rapid weight loss. This program is taught in many books as an acupressure face lift program. Too often, weight loss programs are abandoned because an individual feels that he is not looking better as a result of his efforts. This chapter eliminates this major weight loss saboteur.

Chapter Eleven contains program calendars that will assist the dieter in regaining control over all stress induced weight loss saboteurs presented in *Eliminating the Stress of Weight Loss*. By following a 30-day regimen of self-aided recovery, the dieter will master the recognition and control of their stress induced weight loss saboteurs. Once learned in this manner, the dieter will no longer have the problem of maintaining weight goals. In addition, this chapter is filled with marvelous hints and techniques that have helped dieters for many years.

Addendum #1 - The Biofeedback Reflex Test learned in Chapters Three and Four enables the dieter to detect their Unsuspected Stress Foods with the help of a friend. For those individuals who want to monitor their foods without the aid of a friend, a biofeedback tool, the Jamar® Dynamometer, is introduced. It is an instrument that you can find in most doctor offices. The use of the dynamometer is fascinating and exciting. If you are on friendly terms with your physician, ask him or her to allow you to use their "Jamar® Dynamometer" the next time you visit their office. Since most diet programs limit food choices to 20 or 30 different items, you can arrange a time when you can bring in some foods for testing. By following the simple directions in Addendum #1, you can use this instrument to monitor your Food Instinct Signal as it responds to your personal Unsuspected Stress Foods. The Jamar Dynamometer is available from any local medical supply dealer. An easier and less expensive monitor, "The Food Instinct Monitor" is available from The Monterey Wellness Center. Please refer to page 223.

And finally for those individuals who want additional scientific proof and validation of the efficacy of The Biofeedback Reflex Test, **Addendum #2** presents a clinically reproducible observation. In the scientific community, in order for an experiment to be considered valid, it has to be reproducible by any individual anywhere in the country, using the same equipment and following the same instructions used in the original observation. Addendum #2 uses an instrument found in many hospitals throughout the country. This procedure utilizes computer-generated charts to demonstrate that the introduction of a non-specific food into the digestive pathways of the human body can produce an instantaneous and recordable fluctuation in the strength of body muscles. I encourage and welcome feedback from individuals who clinically reproduce this observation.

How You Will Benefit From This Book

1) The information in this book will help you increase the effectiveness of your present weight loss program by teaching you how to recognize your personal Food Instinct Signal and by exposing The Food Rejectivity Syndrome caused by Unsuspected Stress Foods in your diet. This information allows you to isolate those perfectly good foods in your weight loss regimen that should not be part of your personal program. It is the isolation and rejection of these Unsuspected Stress Foods from a normally good weight loss program that enhances the effectiveness of the program.

2) The information in this book will help you end the unsuspected, and thus unexpected, accumulation of additional fat stores that prevent you from maintaining your hard-earned weight loss goals. You accomplish this by using your biofeedback monitoring system. Like any biofeedback program, the major goal is to enhance your personal awareness of how your body is responding to various stressors in your life. This biofeedback monitoring system enhances your personal awareness of your Food Instinct Signal. Once you gain this awareness, you

will always have it and you will no longer need the biofeedback technique. This signal lets you know when you are about to begin the process of accumulation of fat stores. It is this fore-knowledge that will save you from the weight loss roller coaster.

3) The information in this book will help you end your food cravings and compulsive eating that have sabotaged your weight loss goals in the past. It does this by teaching you the anti-addiction erasure technique that acts to soothe and then to eliminate the sensation of "craving".

4) The information in this book will allow you to enjoy greater health, energy and vitality. It accomplishes this through the removal of the waste by-products of dieting. The removal of these toxic waste by-products represents a major step in the direction of optimizing health and energy levels.

5) The information in this book will allow you more enjoyment of life by showing you how to eliminate old patterns of anxiety, stress and tension related to dieting. It accomplishes this by teaching you how to remove earlier conditioned nega-tive behavioral reaction patterns that are stored in your subcon-scious mind. The information in this section will help lessen the anxiety, stress and tension of all other aspects of your life as well.

6) The information in this book will help you improve your personal beauty by teaching you The Cosmetic Acu-Point Facial Toning Program. This self-treatment works to eliminate lines and wrinkles from your face which will in turn give you a more youthful, healthy appearance.

7) The information in this book will help you improve your self-confidence. This is a result of eliminating old negative behavioral patterns that were preventing you from recognizing and achieving more successful goals in your life.

8) And finally, the information in this book will help you maintain your hard-earned weight loss goals for the rest of your life without having to resort to new weight loss programs every year or so. You will be free from the roller coaster ride of the weight loss syndrome because you will have eliminated your stress induced weight loss saboteurs -- forever.

How To Use This Book

As I mentioned earlier, you may only be affected by one or by several of these weight loss saboteurs caused by your body's natural stress defense mechanisms. To help you determine if a specific chapter relates to your problems, I have highlighted (on both right and left margins of the text) excerpts that I feel represent the most important thought or concept expressed in that section. You can gain a good understanding of the chapter by reading these outer phrases, which I call building blocks. I recommend reading through these building blocks the first time you sit down with this book. Using this reading technique, you can read the entire book in less than one hour. When you find information that you feel is vital to you, take more time to read that page or chapter.

Each chapter represents a compilation of the most significant material pertinent to that specific weight loss saboteur. If a specific saboteur is creating a problem in your life, I list additional references at the beginning of each chapter so you can learn more about the subject matter. Many of the books listed are available at local bookstores. Two excellent catalogs are listed on page 196 for mail order purchase - T.H. Enterprises and Systems D.C.

Although the programs outlined in *Eliminating the Stress of Weight Loss* are designed as self-treatment methods, the dieter is not without support. The Monterey Wellness Center offers audio and visual cassette programs (page 222). We also invite your questions and concerns. You may write or call Dr. Shain (335 El Dorado Street, Monterey CA 93940/ (408) 646-0555). Also, on page 195 there is a directory of organizations listing thousands of physicians and teachers throughout the country who can help you. For your convenience, the address and phone number of each organization is provided. Members of these organizations all understand and can teach the activities explained in this book. They also welcome questions and requests from you.

Also listed are 26 additional national organizations (pages 196, 197 and 198) that can provide information on nutrition, diet and weight loss. Although these organizations may not be familiar with the information discussed in this book, they all provide fascinating, helpful and beneficial information specific to their own interests. Each group is listed with a short explanation of its function along with its address, phone number and contact person. All welcome responses from readers and will answer your questions with caring concern.

Many other excellent books on diet, nutrition and weight control are available. You will find a complete directory of the best and most current books (pages 199 to 202). A glossary (pages 206 to 212) helps you to understand words and phrases used in this book that may be new to you. Also, the index (pages 213 to 217) is designed to help you locate and to reference concepts and ideas presented throughout the text.

All Things Considered

The thrust of the 90's is to take responsibility for your own health and well-being. Self-treatment is a means of assuring control of your life. The extent to which you employ the concepts explained in this book will determine the extent of the results. You will be well rewarded for your efforts.

One thought that will become very clear to you after reading this book is that expecting to lose weight or to even maintain your weight loss goals through restriction of diet alone, is like treating a single symptom of a complex disease.

Although this thought may produce ambivalent feelings, you will be relieved to learn why the results of previous dieting episodes have not been long lasting. You will also be relieved to learn that it was either your lifestyle, stress, heredity or illness that inadvertently triggered your natural stress defense mechanisms and sabotaged your previous weight loss goals.

The principles presented in this book are not meant to be temporary or short term. These techniques, once learned, will benefit you for the rest of your life for they constitute a wonderful health-enhancement program. You will find, as an added benefit, that your health, vitality and strength will improve dramatically and that these improvements will lead to a greater enjoyment of your daily life. Not only will you have a greater sense of well-being, you will have a greater ease in maintaining all of your life's goals.

I encourage you to take a stand. I challenge you to eliminate your stress induced saboteurs — now. Eliminate them now because they are sabotaging the joy, happiness, health and well-being that you deserve in your life.

THE SIX SOLUTIONS
TO
ACHIEVE & MAINTAIN
WEIGHT LOSS GOALS

Use the Art and Science of Biofeedback to
Achieve & Maintain Weight Goals
Eliminate Dieting's Toxic Back-Up Problems
Eliminate Fat Producing Unrecognized Stress Foods
Eliminate Food Addictions Caused by Stress & Tension
Eliminate Conditioned Negative Behavior Reaction Patterns
Eliminate Facial Lines & Wrinkles Caused by Stress & Diet

This chapter gives an overview of the six solutions you need to correct all eight weight loss saboteurs discussed in this book. Each solution represents the self-treatment necessary to erase and eliminate that problem from your life.

Notes:

Thoughts:

Actions:

Questions:

Additional References:

Body Quantum by Fred Alan Wolf
Inner Bridges by Dr. Fritz Fredrick Smith
You Can Heal Your Life by Louise L. Hay
Getting Well Again by Dr. O. Carl Simonton
The Wellness Workbook by Regina Ryan and Dr. John Travis

CHAPTER ONE

The Six Solutions to Achieve
&
Maintain Weight Loss Goals

The First Solution

Use the Art and Science of Biofeedback to Achiev & Maintain Weight Loss Goals

Did you ever wonder how some people can keep their weight down by apparently knowing what to eat and when to eat it? They can automatically control their weight because of an inner sense (intuition) which tells them what is OK to eat. These people have learned to listen to their intuition. This is the secret of becoming and, more importantly, staying thin.

Did you ever say to yourself: "I better not eat this specific food today." or "For some reason, I do not feel like eating this specific food at this time."? If you have, then you were responding to your "Food Instinct Signal". This signal was telling you something and you chose to listen.

This is the secret of successful dieters the world over; whether they know it or not, they have learned to develop and trust in the recognition of this signal.

Everyone possesses this natural instinct that I call "The Food Instinct". This first solution uses biofeedback to teach you to recognize your Food Instinct. You

Use the Art and Science of Biofeedback to achieve and maintain weight loss goals.

You will use biofeedback to discover your Unsuspected Fat Producing Stress Foods.

will learn the Biofeedback Reflex Technique (B.R.T.) in Chapter 4 and 5. This amazing biofeedback technique allows you to recognize your own biofeedback instinct signal. It is this new awareness of your natural "Food Instinct" that will give you greater success in losing unwanted pounds and maintaining your optimum weight.

The Second Solution
Use Biofeedback to Discover Your Unsuspected Fat Producing Stress Foods

The digestion process consistently strives to reduce food to its smallest possible molecules. But, on certain days, because of heredity, lifestyle, stress or a multitude of other factors, your digestive system may find it difficult to break down certain foods into these simple single molecules. Your nervous system recognizes that, at that time, this specific food will only break down into large complex molecules instead of the simple single molecules needed for healthy digestion. These large complex molecules are too big to be used by the body for anything other than emergency storage, i.e., in case of starvation. Normally good food, eaten at this time, puts a lot of stress on your digestive system. If eaten, your body will send it to storage areas called "fat cells".

Your body innately knows this and sounds the alarm, triggering your "Food Instinct" signal.

The Biofeedback Reflex Technique (B.R.T.) will help point out the alarm signal caused by Unsuspected Stress Foods while you are still chewing them — even before you swallow them. The B.R.T. gives you a visual monitor of this neurological response signal.

If you are presently following one of the many successful weight loss programs, you can use the B.R.T. to discover if a specific food in your diet might sabotage your weight loss goals. By recognizing and eliminating Unsuspected Fat Producing Stress Foods from your diet program, your weight loss results will greatly improve.

You will eliminate your food addictions that are sabotaging your weight loss goals.

The Third Solution
*Eliminate Food Addictions
Caused by Anxiety, Stress and Tension*

It is well-known and documented that human beings tend to become addicted to things that are harmful to them such as drugs, alcohol, foods, activities, people, emotions, etc. Most of us are addicted to the very foods that created our weight problems.

Thus, unless you eliminate your personal food addictions, you will crave those foods that are destructive to you and your diet regimen. One day you find that you have lost control. You will gorge yourself silly sending weeks of dieting down the drain.

For years, patients at The Monterey Wellness Center who were addicted to

25

You will eliminate dieting's Toxic Waste Back-up problems that sabotage weight loss goals.

alcohol and drugs were taught a home program to help them subdue the cravings caused by their addictions. With minor adaptations, this self-treatment program effectively conquers food addictions. Whenever a dieter now feels a craving for foods he should not be eating, he can use this sixty-second anti-addiction activity to eliminate those craving "signals".

Fourth Solution

Eliminate Dieting's Toxic Waste Back-Up Problems That Sabotage Weight Loss Goals

Successful loss of weight means that your body is ridding itself of accumulated fat cells plus the debris and toxic substances that are stored in these cells. Depending on the condition of your digestive and elimination system, your body may become a toxic waste dump. The amount of new "stuff" that must be filtered, processed and excreted may increase to a level where your system cannot adequately handle the overage.

The importance of eliminating Toxic Waste Back-Up during weight loss cannot be overlooked for the blockage of "drainage systems" sabotages diet programs.

The goal of this forth solution is twofold: First, the techniques presented energize and revitalize those areas of the digestive system that are put under additional stress during weight loss. Secondly, the techniques drain and flush out the waste products created during weight loss.

The Fifth Solution
Eliminate Negative Behavior Reaction Patterns Created by Prior Conditioning

You're twelve years old and you weigh 150 pounds. You're short and being taunted by peers and parents because of your weight problem. This teasing ingrains a destructive, poor self-image into your young, nieve psyche. No matter what you do with your weight in future years, you will still have the problem of seeing yourself as a "fat little child". Even a pretty and trim young person, taunted about weight by siblings or even by a jealous parent, may retain a negative self-image that continues to sabotage later adult desires.

Is it any wonder that this person has difficulty losing weight? Is it any wonder that this person regains lost weight? No weight loss program will ever give this person permanent results until she learns to create a new, more positive self- image. It is subconscious negative behavior reaction patterns, based on old pictures, that sabotage weight loss programs.

This fifth solution helps you eliminate destructive self-image concepts stored in your subconscious memories. Creating a new self-image to replace old negative behavior reaction patterns also helps eliminate many other present day negative behavior activities.

You will eliminate old negative behavior reaction patterns created by prior conditioning.

The principles presented are not meant to be temporary or short term. Once learned, these techniques benefit dieters for the rest of their lives.

Sixth Solution

Eliminate Facial Lines & Wrinkles Caused By Stress & Diet

Solution Six presents the Cosmetic Acu-Point Facial Toning Program of The Monterey Wellness Center. This activity will eliminate facial lines and wrinkles that have been created through anxiety, stress, tension and rapid weight loss.

Solution Six also gives you techniques to improve facial skin complexion, tone and color. These techniques have been gleaned from some of the most expensive health and beauty spas in the world. They are very effective methods that will enhance the self-confidence of any dieter. These weight loss spas have known that when you create a new body through weight loss, the results will be maintained forever if you also improve the dieter's self-confidence and facial beauty.

CHAPTER TWO

LIFE'S DAILY STRESSORS SABOTAGE WEIGHT LOSS GOALS

UNDERSTANDING THE STRESS OF WEIGHT LOSS

The Stress of Weight Loss
A Perfect All-or-Nothing Response
The Relationship to Dieting
Unsuspected Stress Foods
Why Weight Loss Programs Fail
Addictions and Your Response Pathway Mechanisms
Addicted to Your Unsuspected Fat Producing Stress Foods
How Stress, Lifestyle or Illness
Trigger Unconscious Negative Eating Reactions
Overcoming the Negative Factors Influencing Weight Loss

This chapter discusses the stress mechanisms that sabotage weight loss goals. You will learn how heredity, lifestyle, stress and illness can all create negative reaction pathways within your nervous system. Once created, these reaction pathways operate on automatic pilot -- without conscious control. It discusses the mechanisms by which you become addicted to specific foods. Understanding how your life's daily stressors affect dieting starts the process to free you from the weight loss roller coaster. These reaction pathways can also sabotage other aspects of your life making your days just a little harder than they really need to be.

Notes:

Thoughts:

Actions:

Questions:

Additional References:

The Stress of Life by Dr. Hans Selye
Stress, Sanity and Survival by Robert Woolfolk
Nutrition and Your Body by Benjamin and Sarah Colimore
Behavioral Kinesiology and the Autonomic Nervous System by
    Dr. John Diamond
The Physiopathology and Treatment of Functional Disorders by
    Dr. George Whatmore and David Kohli

CHAPTER TWO

How Your Life's Daily Stressors Sabotage Weight Loss Goals

Understanding The Stress of Weight Loss

"Unrecognized and retained stress in your nervous system is the cause of much dysfunction. This unrelieved stress can distort the self-regulatory systems, affecting digestion, tissue maintenance, cell rebuilding, and other important health maintaining functions."

Dr. Michael Lessor, M.D.
Nutrition and Vitamin Therapy

The Stress of Weight Loss

When your body must respond to any "Stressor" in your daily life, an immediate, measurable surge of energy is produced to facilitate all of the automatic body responses necessary to make a response. This surge can be seen and recorded using electromyograph equipment (Figures 2.1 and 2.3). Your central nervous system will direct this surge of energy to those parts of the body where it is most needed. This natural energy, the "Energy Reservoir", works like a built-in power-pack to help maintain your energy at constant comfortable levels. From it you draw the additional energy needed to successfully confront any stressful situation in your daily life.

This natural energy, the "Energy Reservoir", works like a built-in power-pack to help maintain your energy at constant comfortable levels.

31

Your physical responses triggered by this directed surge are created by the Response Pathway Mechanism.

The pathway for this energy channel is called the Response Pathway. The physical responses triggered by this directed surge are created by the Response Pathway Mechanism (Figure 2.2). The following story illustrates this Mechanism.

While visiting Hollywood, let's say you are invited to attend the taping of an extremely famous quiz show, "The One Million Dollar Question". While you are quietly sitting in the studio audience, the producer of the show comes up to you and says, "Our contestant for today is ill and we have been watching you on our monitors. We would like you to take his place and try to win one million dollars today."

These words, no matter how softly spoken, will trigger all of your automatic response pathway mechanisms. Before the word "yes" has a chance to form in your mouth, the following automatic response reactions have occurred in your body.

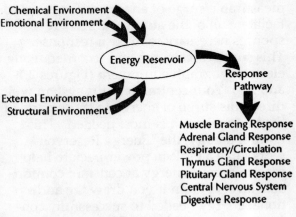

Chemical Environment
Emotional Environment
Energy Reservoir
Response Pathway
External Environment
Structural Environment

Muscle Bracing Response
Adrenal Gland Response
Respiratory/Circulation
Thymus Gland Response
Pituitary Gland Response
Central Nervous System
Digestive Response

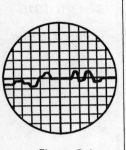

Figure 2.1
Before Stimulation

Figure 2.2
The Response Pathway Mechanism

Response #1 - The Muscle Response

The tension in your body musculature, and, as a result, your posture, will become altered. Your muscles will brace, preparing your body to meet the challenge. You will be taking a stance, putting up your guard, putting on your muscular armor. This is a posture of strength and control that gives you the sensation of being in control of the situation.

Response #2 - Endocrine System Response

The adrenal glands produce and control the rapid release and distribution of a multitude of hormones and chemicals, including adrenalin and cortisone. They also affect your body's balance between the chemicals sodium, potassium and chloride. These auto- matic chemical reactions then initiate the following:

- Because of the release of large amounts of cortisone which now circulates through-out your body, sensations of pain are eliminated. Any new scratches or cuts that you might receive during the filming of the show would not be felt until hours later when your adrenal glands calm down.

- Fatigue and exhaustion disappear as your body is revitalized by the production of greater quantities of natural drugs and hormones. The strength and vitality that you thought would take an entire night's sleep to recharge is instantly yours.

Response #3 - The Respiratory Response

The rate and pattern of your breathing is altered. Prior to this response your breathing may have been shallow and limited. The release of adrenalin will serve to dilate the bronchi in your lungs to allow a maxi-

Fatigue and exhaustion disappear as your body is revitalized by the production of greater quantities of natural drugs and hormones.

Figure 2.3
After Stimulation

This biological drug will give you a euphoric feeling that allows you to believe that you will be extremely successful.

mum intake of oxygen. Your respiration rate will increase so that your body will become better oxygenated. Your brain and associated thinking capabilities can now respond more successfully.

Response #4 - The Circulatory Response

The blood vessels servicing the surface areas of your body constrict. Your hands and feet may feel much colder because less blood can get to these areas to warm you. But, your brain and other vital organs now have a more than adequate supply of nutrients and a good drainage of waste material. And, since the excess cortisone (Response #2) will prevent you from feeling scrapes and cuts that you may suffer, this constriction of the blood vessels will prevent you from unknowingly bleeding all over the TV studio.

Response #5 - The Immune Response

Your immune system produces and controls the release of those substances (antibodies, leukocytes, lymphocytes, etc.) which are responsible for combating any infections in the body. It also removes any foreign or waste debris. When this system is activated, you are no longer bothered by that sore throat and runny nose you had all day long. A side benefit is that the small infection that has been trying to initiate itself in a small finger cut will also be attacked by this immunological response and the cut will heal faster.

Response # 6 - The Psychological Response

The pituitary gland is responsible for many physiologic and psychologic changes in your body. A primary function is its production of endorphins. This biological drug will give you a euphoric feeling that allows you to believe that you will be extremely successful. This gives you the psychological boost of truly believing.... "I know I can . . . I know I can . . . I know I can", thus allowing you to compete with a very positive attitude.

Response #7 - The Digestive Response

Just 30 minutes before going to the studio, you ate lunch. The food has been undergoing the slow process of digestion in your stomach. But now, the automatic response pathway mechanism will secrete large amounts of additional hydrochloric acid (a natural digestive enzyme) into the stomach to immediately emulsify any food present. This rapidly processed food is then quickly passed out of your stomach. Your body is now free of an otherwise major distraction.

Response #8 - Automatic Pilot Response

A physical shift in bodily control takes place in which the sympathetic nervous system (conscious decisions and voluntary controls) becomes secondary to the para-sympathetic system which better allows you to respond "automatically" to "gut" instinct responses.

Response # 9 - The Sexual Response

The interaction of all of the chemicals and hormones that are being dumped into your system has the net effect of turning off those mechanisms that will initiate any sexual desires. When trying to think of the correct answer you will not be distracted by that attractive person in the front row.

This physical Response Pathway Mechanism is very successful at helping you survive.

35

*Until
it is no
longer
perceived
as a threat,
the
isolation
and
rejection
of the
foreign
element
is the
major
goal
of this
mechanism.*

A Perfect All-or-Nothing Response

The most interesting fact about this Response Pathway Mechanism is that there is never a partial triggering. Your nervous system will always exact a complete firing of all response pathway systems. All systems are triggered together. For instance, the bite of a poisonous snake will cause the combined responses to be triggered in this sequence: The poisonous chemical (snake venom) enters the body as a "foreign element". The immune system acts to send out scavenger cells throughout the body to isolate and remove the invading chemical. The adrenal and pituitary glands respond to give you the power and the will to do anything necessary to stay alive. Your respiration, digestion, circulation and other bodily functions will be automatically altered to aid in your fight for survival. Your body energy shifts to conserve energy wherever it can, while directing more power to survival. This physical Response Pathway Mechanism is very successful at helping you survive a snake bite. The isolation and rejection of the foreign element is the major goal of this mechanism until it (the foreign element) is no longer perceived as a threat.

The Relationship to Dieting

This very same sequence of systematic responses that is called into play when your life is threatened, is triggered when you eat a poisoned food. This same procedure ensures survival in case of accidental food poisoning. How well you fare depends not on the strength of the poison, but rather on your body's ability to respond adequately.

As a further aid to understanding how all this applies to dieting, let's examine how your body processes food and converts it into fuel. When you eat a piece of meat, your body simplifies this giant nucleo-protein into its component parts — nucleic acid molecules and protein molecules (Figure 2.4 from (*Nutrition and Your Body* by B. Colimore).

The large nucleic acid molecules are then reduced to their basic parts: ribose and nitrogen bases. The proteins are also broken down into their simplest component parts which are the amino acids. These basic elements then become the building blocks through which you build new tissue for both growth and repair.

How well you fare depends on your body's ability to respond adequately.

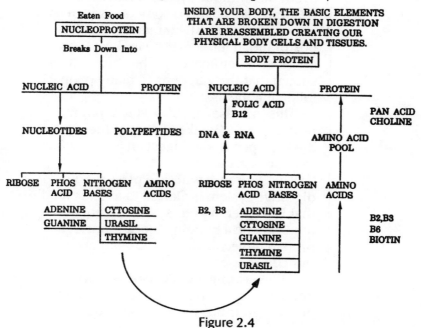

Figure 2.4
Processing Food into Fuel (for energy) and Cells (for repair and growth).

This response will also take place if your digestive system is alerted to a "toxic food".

Your new building blocks are transported from the intestines to collection and distribution areas where we can imagine a large genetic pool to exist. Genetic blueprints (RNA) are free to take from this pool the correct amount and ratio of elements needed to create new body protein. This body protein then becomes your cells which then become blood, muscle, ligaments, bone and so forth.

Unsuspected Stress Foods

An interesting note is that for the past 30 years, physicians who have studied human blood chemistries have observed that the amount of circulating leukocytes (from the immune system's reaction - Response #5, page 34) sometimes increases after eating. This phenomenon is called postprandial (after eating) leukocytosis (increase in leukocytes). The studies have shown that this happens to many people following meals.

This increase in leukocytes is not the reaction to the intake of proper fuel. But, it does happen if you ingest an even slightly toxic substance.

This response will take place if your digestive system is alerted to a "toxic food" that has entered your system -- not as a food chemical building block -- but as a protein molecule that is too complex or difficult to use.

When eating an Unsuspected Stress Food, instead of responding to "building blocks", your body will respond as if you had just eaten a mild "poison" (as in our example of the snake bite). All systems of your response pathway will be triggered (Figure 2.2). And, as you have seen, in addition to the immune system shifting into gear, all of the other physiological body responses are triggered.

Why is This Important?

Many people have lost sight of the most important aspect of eating -- the replenishment of life. "Convienience" is now the deciding factor in choosing a meal. "If you can eat it without dying then it has to be O.K." People of an industrial society, are eating an "industrial diet". The unfortunate prerequisites for an edible substance today are: traditional ethnic acceptance, pleasing appearance, easy availability, quick edibility and peer pressure. Taste does not necessarily have to be a factor, as proven by food trends over the centuries. All edible substances cause a chemical reaction to take place within your body. Let's explore that reaction.

Why Weight Loss Programs Fail

Now it's time to consider some important questions concerning weight-loss diets:

Why does the excess weight return after ending a successful weight loss program?

Why does one diet work well for some people, but not for others?

Thus, when eating Unsuspected Stress Foods your body will, instead of responding to "building blocks", respond as if you had eaten a mild "poison".

A successful weight loss program must include the ability to ascertain which foods your body is using as fuel, which are causing physical distress and which are being stored as fat.

Why do some people actually start to feel weak and ill when they try one specific diet while others will feel exceptionally good?

Why will the diet that allowed your friend to lose weight, cause you to gain weight?

The answers to these questions can be found in the automatic Response Pathway Mechanism. How or what you eat is not as important as how your body processes the food you eat. If, as you have seen, your body cannot process a food into effective building blocks, it will do one of two things:

Body Response #1 - Recognize it as an undigestable molecule and store it because it is too difficult to utilize at the moment. **It is these stored complex molecules that constitute the fatty deposits of the body's fat reservoirs.**

Body Response #2 - Attack it like a poison, causing a full-blown Response Pathway Mechanism reaction. **It is this reaction, as we will learn in the next section, that creates so many of our food addictions.**

A successful weight loss program must include the ability to ascertain which foods your body is using as fuel, which are causing physical distress and which are being stored as fat.

Addictions and Your Response Pathway Mechanisms

Now consider this second story:

A young child enters the school cafeteria. He picks up his tray and silverware and proceeds to the line where he will be served. The head nutritionist, a particularly nasty person, does not like this child. Just as the child reaches to get his food, the nutritionist slaps the child on the wrist. The next day, the child is slapped again.

If this slapping is repeated, by the third or fourth day the child will be leary before he extends his hand to get the food. He will brace his muscles to prepare for the expected slap. He will tighten the muscles of his arms, neck, shoulders and back. He may prepare to run by tightening his leg muscles.

The longer this situation persists, the more this child will be affected. Soon his entire stress defense mechanism will respond. He will brace against the slap; his muscles will tighten; his adrenal, thymus and pituitary glands will be stimulated; his digestive system will respond; and his electrical patterning will shift.

As the abuse continues, the child may start reacting against the attack even before he picks up his tray. He may eventually begin to react before walking into the cafeteria, possibly while still in the classroom prior to the lunch break. And if the situation still persists, the child may unconsciously call his entire stress response mechanism into play at the mere thought of eating lunch.

The child is being conditioned. If nothing is done to eliminate this conditioning, a pattern will develop.

It is this "anxiety" alone which can now continually and needlessly trigger his body's Response Pathway Mechanisms.

The longer this situation persists, the more ingrained, automatic and conditioned the response pattern becomes. The external stimulus of the slap on the wrist has produced an entirely physical response from the brain and automatic nervous system via the body's defense pathway mechanisms. The reponse now begins to produce anxiety. At some point the child may start voicing some complaints whenever it is time to eat. The back of his neck and shoulders are tight; his neck hurts; his stomach hurts; he feels tired; he says that he does not feel like eating. He will also develop behavioral problems related to trying to avoid going to the lunchroom.

The child is being conditioned. If nothing is done to eliminate this conditioning, a pattern will develop. At this point, we can even remove the causal agent and the conditioning will continue. This means that even after the parents move the child to another school, he will still brace his muscles and become ill at ease every time the thought of eating lunch occurs to him.

He has developed an anxiety (fear and worry projected into the future) about eating lunch. It is this "anxiety" alone which continually and needlessly triggers his body's Response Pathway Mechanisms -- even when the original stimulus is removed.

From this hypothetical story, we can see how our bodies retain a physiological response long after the situation that caused it has passed. Without being aware that we are doing so, we "worry" or continue to think about past stressful events.

At times, based on previous experience, we conjure up a possible future reaction to these threatening events. When we do this, the body's automatic Response Pathway Mechanism responds even though there may be no action to take. In this manner, a cycle can develop where a potentially exhausting physical response is continually initiated as a result of the unconscious memory of a conditioned response.

Yet, properly activated, this physical response is a very beneficial mechanism.

For example: When a beginning typist sees a given letter of the alphabet, her entire cognitive picture of the exact finger to be used, the location of the key to be pressed, and all that her instructor has told her about that letter, slowly evokes the correct response so that the proper key is pressed. After much repetition and practice, the sight of the letter elicits the key pressing response directly, without her intervening act of consciously trying to figure it out. Eventually, just the sight of an entire word may be enough to automatically trigger the correct sequence of key pressing.

Another example of a protective automatic motor response is seen in the experienced soldier who at the slightest unknown sound dives for shelter with minimal conscious thought and thus minimal delay. The automatic response was acquired from previous conditioning experiences during which the correct response became automatic through repetition. Once under cover he has the time to assess his position and possible actions.

This type of conditioning enables us to acquire automatic motor responses, including complex motor skills, that conserve time and energy.

43

This feedback pathway, set up from your response pathway organ systems, becomes your Conditioned Response Pathway.

This type of conditioning enables us to acquire complex automatic motor responses that conserve time and energy. These automatic motor responses occur without the aid of conscious thought and with very little attention. This is because the repetition of an action or event creates a Conditioned Response Pathway (Figure 2.5) that tells the central nervous system if that complex motor response was adequate, if the response worked correctly and if an additional response is needed.

Addicted to Unsuspected Fat Producing Stress Foods

A conditioned reflex response, created through repetition, establishes a link back to the central nervous system from the response organs themselves. It is this

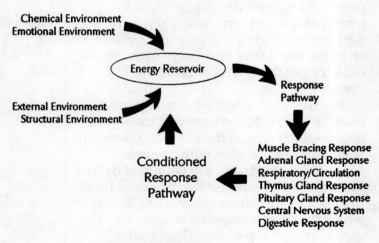

Figure 2.5
The Conditioned Response Pathway Mechanism

feedback pathway link, set up from the response organs, that is the Conditioned Response Pathway. It is this automatic feedback signal from the organ systems which confirms the completion of the correct response reflex. This feedback pathway, set up from your response pathway organ systems, becomes your Conditioned Response Pathway. Every conditioned action (the child's slap, the typist's response to the keys, or the soldier's response to the unknown noise) has its own Conditioned Response Pathway.

The relationship between your Unsuspected Stress Foods and food addictions will be discussed in detail in the next chapter. But for now it is important to recognize that when you eat any food that your central nervous system interprets as a stress food, your entire Response Pathway Mechanism will be triggered. As we have seen, this is the mechanism that helps assure your survival in case of mild accidental poisoning.

If one of the foods in the dieter's previously normal choices of foods was an unrecognized dietary distress agent, it caused a triggering of his body's Response Pathway Mechanism each time the food was eaten. Most times, this happened without the dieter being aware of the mechanism. The persistent triggering of this mechanism by a specific stress food continually initiates the same firing sequence in response to that "trigger". And, just as we have seen with the child's slap, the typist's word response and the soldier's response to the unknown noise, it is exactly

It is this Conditioned Response Pathway Mechanism of Unrecognized Stress Foods that can sabotage the best of dieter's intentions.

Behavioral actions created through prior conditioning that are triggered by present day internal exhaustion signals, have a detrimental effect on weight loss goals.

this type of continued repetition that will create a conditioned (non-thinking) response.

It is this Conditioned Response Pathway created by an Unsuspected Stress Foods that can sabotage the best of dieter's intentions. Ironically, when you eliminate this particular food from your diet for weight loss reasons, your body begins to crave the food as if it were an addictive substance such as alcohol or drugs. This explains why a dieter may one day thoughtlessly gorge herself on all of the worst possible foods after weeks of successful dieting. This one action caused by the Conditioned Response Pathway's cravings for an addictive food destroys weeks of dieting resolve.

How Stress, Lifestyle or Illness Trigger Unconscious Negative Eating Reactions

During periods of high stress, excessive lifestyle activity or illness, one or more of the response pathway organs (Figure 2.2) can become exhausted. As exhaustion continues, the organ can begin to struggle to maintain its ability to respond correctly. Other body systems realize that at this time that organ may need an additional boost from the central nervous system. But before this happens, a signal must be sent to the central nervous system from the exhausted organ.

As we have seen, each of the six response organs has many conditioned response pathways already in place which go directly back to the central nervous system. Although these specific conditioned pathways have been set up to carry a specific action response, these pathways are convenient routes to carry the new fatigue/stress signals. Thus the fatigued organ can choose from any of the many Conditioned Response Pathways already in place to send its own exhaustion signal.

Problems result when these internal exhaustion signals accidentally initiate the conditioned behavioral action response already established in that chosen pathway. This means that some behavioral pattern normally associated with an earlier life situation is accidentally initiated solely by the alarm signal of an exhausted, stressed-out organ. Behavioral actions created through prior conditioning that are triggered by present-day internal exhaustion signals, have a detrimental effect on weight loss goals.

Your weight loss goals are sabotaged when fatigue or exhaustion signals trigger negative behavioral reaction patterns that have been stored in your automatic mechanisms. You can, without conscious thought or desire, find yourself using negative behavioral responses and actions that you thought you had overcome years ago. This phenomenon accounts for many of those unconscious negative behavioral actions you revert to when you are "stressed out".

Also, it is the answer to those people who ask: "What made me ruin my diet by ordering and eating all of the wrong food at lunch?" You will get more information

You can, without conscious thought or desire, find yourself using negative behavioral responses and actions that you thought you had overcome years ago.

*Recognizing
and then
eliminating
personal
food
stressors
is the first
of six
significant
ways of
enhancing
the
effectiveness
of all
weight loss
programs.*

on this in Chapter Nine which demonstrates how to control those subconscious negative behavior reaction patterns that are triggered by stress, lifestyle or illness.

Overcoming the Negative Influences of Weight Loss

A successful weight loss program must include the ability to ascertain which foods the dieter's body is using as fuel, which foods are their personal stress foods and which foods are being stored as fat.

Chapters Three and Four will show you how to recognize your central nervous system's Food Instinct Signal which is sounded in response to Unsuspected Fat Producing Stress Foods. Acquiring the awareness of this signal will allow the dieter to successfully recognize and thus avoid these weight loss saboteurs. Chapter Six then demonstrates how to eliminate food addiction cravings generated from this Conditioned Response Pathway Mechanism.

A major tool for maintaining successful weight loss will come from your ability to recognize and thus to develop your awareness of your Food Instinct Signal. It is your new awareness of this signal that will completely revolutionize your concept of dieting. As you will learn, it is the recognition and elimination of personal food stressors that is the first of six significant ways for you to enhance the effectiveness of your current weight loss program. It is also the means by which you will maintain successful weight goals after achieving them.

ELIMINATING FOOD REJECTIVITY FROM YOUR WEIGHT LOSS DIET

UNSUSPECTED STRESS FOODS SABOTAGE WEIGHT GOALS

The Food Instinct Signal
Food Rejectivity
Dietary Distress
The Poisoning of America
Food Allergies & Weight Problems
The Food Rejectivity Syndrome
Addicted to the Chemicals in Your Food

This chapter introduces you to the Food Instinct Signal. This is a reflex signal that exists within your automatic nervous system. Some people call it their "intuition". The discovery that your Food Instinct Signal is triggered by Unsuspected Stress Foods represents a major advancement in the understanding of the weight loss roller coaster. This chapter explains the Food Rejectivity Syndrome and how Unsuspected Stress Foods sabotage even the best of weight loss programs.

Notes:

Thoughts:

Actions:

Questions:

Additional References:

How to Survive the Poisoning of America by Dr. Joel S. Shain
Stress and the Art of Biofeedback by Dr. B.B. Brown
Food Additives - A Consumer's Dictionary by Ruth Winter
Handbook of Nutritive Value of Processed Food by M. Rechcigl
Holiday Dinner Menu by the American Council on Science and Health

CHAPTER THREE

Eliminating Food Rejectivity From Your Weight Loss Diet

Unsuspected Stress Foods Sabotage Weight Loss Goals

"The accumulation of body fat results from your body's attempt to store Unsuspected Stress Foods until lifestyle, stress or health factors allow for better digestion."

Dr. Joel S. Shain
Life Empowerment

The Food Instinct Signal

Wild animals are equipped by nature with an unconscious mechanism that gives them the awareness of what is and what is not edible. I call this built-in awareness of what to eat, the Food Instinct. When a wild animal is sick, it will instinctively eat those foods which will aid healing and avoid those foods which may inhibit healing.

The human animal is equipped with this instinct, also. It is the awareness of this instinct that may stop you from eating a particular food which you normally enjoy after tasting it because it does not "feel" right at that particular moment. Whether or not the food is eaten depends upon how much faith you have in your awareness of this "feeling".

Whether the food is eaten or not depends upon how much faith you have in your awareness of this "feeling".

When this biofeedback signal is strong enough, most people recognize this internal response to "stress foods".

Experiments show that babies up to the age of one year choose healthy, natural foods when given a choice of diet. Without fail, babies choose foods which are naturally best for their well-being. By the age of two years however, the child has already become conditioned and socialized so that this natural food instinct is no longer considered in making food choices.

Such experiments prompted the following questions:

By applying this same natural instinct, are some people able to control their weight by knowing what to eat and when to eat it?

Can dieters automatically either control or maintain weight through use of an inner sense or intuition that tells them what is OK to eat?

Can anyone learn the secret of controlling their weight by just listening to their intuition?

Research results show that the answer to all of these questions is "Yes".

This is the secret of successful dieters the world over. They have developed and learned to trust their recognition of this signal. What is amazing is that many are not aware of how they are using this signal to control their weight.

Everyone possesses this biofeedback signal and when it is strong enough, most people recognize this internal response to foods that trigger the "response mechanisms". At times, after tasting a specific food, you may have said to yourself, "I better not eat this food today." or "For some reason, I just don't feel like eating this food at this time." In these instances you were responding to your Food Instinct Signal as it interprets the response mechanisms.

By developing an awareness of this signal, dieters can learn to recognize specific foods which, if eaten at that particular moment, will be stored in those reservoirs we call fat cells.

Food Rejectivity

Food Rejectivity takes place when a food-substance which is eaten initiates a stress reaction from the body in response to that food. This reaction - a function of the stress response mechanism - causes all systems of the Response Pathway Mechanism to fire. It is important to remember that Food Rejectivity responses are directly related to the dieter's lifestyle, stress, heredity and degree of illness. Since these factors constantly shift, the body's ability to successfully process and utilize a specific food also varies. A food that is fine for the dieter on one day may be a dietary distress agent on another. Therefore, it is important for the dieter to develop the ability to recognize body reflex (biofeedback) responses to food, on a daily basis.

In Chapter Four, you will learn a technique that allows you, the dieter, to observe and develop your "Food Instinct Signal". Recognizing this natural biofeedback signal warns the dieter of Food Rejectivity responses.

If a dieter learns to develop an awareness of their Food Instinct, it will point out those foods that will be stored as fat cells.

*These large
complex
molecules
are too
big and
too
difficult
for the
body
to use for
energy,
repair or
reproduction.*

Dietary Distress - The First Category of Unsuspected Fat Producing Stress Foods (Body Response #1 - Page 40)

The purpose of eating is to convert food into fuel for life and energy. Normally the act of eating and digestion is intended to reduce food to its smallest possible molecules (Figure 2.4). It is these basic molecules that are used by the body to produce new cells, to repair old ones, and ultimately to give strength and energy.

On any day, because of heredity, lifestyle, stress or illness, your digestive system may find it difficult to break down one of your diet foods, into these simple single molecules. The Food Instinct Signal automatically "fires" when your central nervous system recognizes this Unsuspected Stress Food -- one which will allow large complex molecules to enter the digestive system instead of the more usable simple single molecules. These large complex molecules are too big and too difficult for the body to use for energy, repair or reproduction. These molecules are then stored in those storage reservoirs we call "fat cells" and saved for emergency use. Because these fat deposits are difficult to use, the body will draw on them only when no other choices are available.

In this category there are 3 sub-groups:

1) High Stress Periods

Eating food at the wrong time or under less than advantageous conditions can result in poor digestion. When the dieter is undergoing a great deal of stress, it is normal for the body's survival mechanism to be concerned with ensuring survival, not enabling digestion. Therefore, digestive systems can temporarily work in a very inferior manner. During stress periods, even good foods will not be digested well. These then difficult-to- use food components will be treated as foreign elements. The body's in ate processes know this and trigger the Food Instinct Signal.

2) Heredity and Illness

These factors affect people with some underlying problem in one or more of their digestive organs. For example, a poorly functioning gall bladder produces less bile, thus giving an incomplete breakdown of fatty foods into building blocks. The percentage of poorly digested fatty molecules depends upon the degree of dysfunction of the gall bladder (varying from 99% function down to 1% function). Another example would be a poor supply of digestive enzymes in the stomach resulting in incomplete breakdown of proteins. The incomplete breakdown of any food to its basic molecules will always trigger your Food Instinct Signal because poorly digested foods tend to be stored in reservoirs for future use.

The incomplete breakdown of any food to its basic molecules will always trigger your Food Instinct Signal.

The National Research Council (NRC), considers 8,627 food additives and 3,410 cosmetic ingredients as "substances of possible concern".

3) Lifestyle - Energy Distress

What the body can tolerate one day, may react like a food stressor on another day. For instance, extremes in lifestyle activity can cause an individual to experience days when they feel "exhausted" or "wiped-out". On these days, that person will find that a normally health-giving food may require more energy to digest than the body will receive from it as fuel. He will have a net loss of energy as opposed to a gain. A food that will rejuvenate you when you have "good" energy may make you feel weak or spaced out on a day when your energy levels are low. This is another way in which a food that will be burned as fuel one day may be stored in fat reservoirs on another day. Your Food Instinct Signal will always fire if your body has difficulty processing a specific food.

The Poisoning of America * - The Second Category of Unsuspected Stress Foods (Body Response #2, p. 40)

The National Research Council (NRC), considers 8,627 food additives and 3,410 cosmetic ingredients as "substances of possible concern". After exhaustive studies, the NRC concluded that no toxicity information was available for 46% of these substances and that for only 5% (of the remaining 54%) did there exist a complete health hazard assessment. And even these health hazard assessments only tested for cancer- producing tendencies. They did

*From Dr. Shain's book, *"How to Survive the Poisoning of America"*

not account for the full range of micro-changes taking place within test animals. The NRC concluded that the food industry should be conducting long term studies in the areas of neuro-toxicity, allergies, inhalation studies and chronic (long-term) dys-function studies. There are presently no tests available to determine if a food additive is causing toxic reactions in the brain or nervous system, even though a great many scientists believe that neuro-toxins are a greater problem then carcinogens.

In America today, any one food can contain hundreds of the 8,627 food additives known to the industry. There are even 300 "standard" foods, including ice cream, catchup and mayonnaise -- foods that we take for granted -- that do not even need to list the chemical additives used just to process them. For example, ice cream can have 30 different additives. Foods are dyed to change the way they look; preserved to ensure long shelf life; flavored with any of the 2,000 flavoring additives; injected with growth hormones, ripeners, metabolic agents, emulsifiers, texturizers, stabilizers and sweeteners. Many of these "every-day" additives are known carcinogens. Today's supermarkets sell foods with warning labels, "This food contains known cancer-producing chemicals."

As if this were not bad enough, even our non-processed fruit and vegetable products are contaminated with poisonous pesticides of controversial, untested and thus unknown complications to the human body.

There are presently no tests available to determine if a food additive is causing toxic reactions in the brain or nervous system even though a great many scientists believe that neuro-toxins are a greater problem then carcinogens.

In America today, over 60 million people suffer a wide range of allergenic reactions to chemical additives and pesticides.

Food Allergies & Weight Problems

What happens to the human body that digests even the most minute quantities of thousands of different additives and chemicals on a daily basis? In America today, over 60 million people suffer a wide range of allergenic reactions to chemical additives and pesticides ranging from, but not limited to: headaches, fatigue, dizziness, joint pains, body and organ discomforts, irritability, depression, confusion, blurry vision, digestive and elimination problems, irregularites in heart beat and blood pressure and even menstral irregularities. The list expands through the entire range of allergy reactions -- and even death.

As described in Chapter Two, if a particular chemical/additive triggers the Response Pathway Mechanism (and thus the Food Instinct Signal), that substance will be treated by the body as a "foreign element" or even a "poison". This system will be triggered each time that chemical is ingested. If any food in the dieter's restricted diet contains a stress triggering food additive, natural carcinogen or other substance that triggers these defense mechanisms, the body will process that food as a "foreign element" (pages 36 - 38) and store these "foods" in reservoirs we call fat cells. The dieter will stop losing weight and may even start gaining.

Thus, unknowingly and quite uncontrollably, weight loss goals are destroyed. As a further insult, weight loss goals are further sabotaged when the dieter starts losing weight -- i.e. the destruction of stored fat cells. Rapid destruction of fat cells releases large quantities of these stored toxins into the system at one time. This causes further toxicity discomforts (Chapter Eight) which continue to undermine even the best of weight loss regimens.

Whether you are one of the 60 million people affected and HOW you are affected depends on your lifestyle, stress, heredity and degree of illness at any given time. What is known for sure is that if any chemical substance in the dieter's restricted diet represents a dietary distress agent, it will trigger the dieter's Food Instinct Signal every time it is eaten. Using the Biofeedback Reflex Test (B.R.T. described on page 73 and 81), dieters will be able to observe their Food Instinct response to foods that are difficult for their body to process. This recognition of personal Unsuspected Stress Foods allows dieters greater success in controlling and maintaining weight loss goals.

The Food Rejectivity Syndrome --
Chemicals Sabotage Weight Loss Goals

When any of your personal Unsuspected Stress Foods are eaten, your body's Response Pathway Mechanism is triggered (Figure 2.2) to help assure your survival rather than assuring good digestion.

What happens if you continue to eat a food that your body considers a "foreign substance - a poison"? Tests have shown

If any chemical substance in the dieter's restricted diet represents a dietary distress agent, it will trigger the dieter's Food Instinct Signal every time it is eaten.

What will happen to your Response Pathway Mechanism if you continue to eat a food that your body considers a "foreign substance"?

that persistent triggering by a specific stress food can cause functional distortion to any of the seven Response Pathway Systems (page 32). Depending upon your lifestyle, stress factors, heredity and degree of illness plus the length of time you have been ingesting these chemicals, you can experience symptoms ranging from micro (undetectable) up to severe difficulty within any of the following systems (from page 33);

Response #1 - The Muscle Response
Continual micro-bracing of body muscula-ture can eventually produce tension, spasms and pain in the muscles of the neck, shoulders, mid-and-low back. This type of muscle pain then produces another series of complaints such as headache, joint pain, fatigue and other body discomforts.

Response #2 -Endocrine System Response
Continual micro-stimulation of this system produces fatigue, exhaustion, "everyday discomforts", enhanced sensitivity to pain, respiratory disorders and degrees of insom-nia.

Response #3 - Respiratory Response
Continual micro-stimulation of this system causes irregular breathing patterns which effect oxygen intake. Inferior oxygenation of blood causes dysfunctions such as mental confusion, dizziness, lightheaded-ness and other micro-trauma to all body cells dependent on oxygen supply.

Response #4 - The Circulatory Response
Continual micro-trauma to this system causes poor circulation, cold hands and feet, headaches, "migraine" headaches and cardio-vascular difficulties.

Response #5 - Immune Response
Continual micro-trauma to this system causes decreased immunity which leads to frequent colds, flu, infections and various discomforts such as sneezing, coughing, running eyes, etc.

Response #6 -The Psychological Response
The continual micro-trauma to this system causes irritability, depression, anxiety and varying degrees of neurotic behavior.

Response #7 - Digestive Response
Continual micro-trauma to this system causes difficult digestion, abdominal pains, excess gas and elimination problems.

Although this listing is only partial in nature, every reader, at one time or another, has experienced any number of these symptoms and questioned whether one of the foods in their diet created that problem.

Dieters by choice, limit their intake of food to certain types of "good" foods. If one of the foods in the dieter's weight loss program is an Unsuspected Stress Food, the dieter will not feel as good as he did before starting the diet. **This represents the first major Dietary Distress Saboteur.** The recognition and elimination of stress foods from a weight loss diet will ensure a healthier, safer and more successful weight loss program.

Addicted to Chemicals in your Foods

The most unfortunate aspect about Food Rejectivity is that an individual experiences this dietary distress as a pleasurable sensation. The ingestion of the "stressor", in this

Every reader, at one time or another, has experienced any number of these symptoms and questioned whether one of the foods in their diet created the problem.

If a food that triggers this pleasurable mechanism is omitted from the dieter's weight loss regimen, it will be only a matter of time before the dieter begins to "crave" that food.

case the Unsuspected Stress Food, causes the pituitary gland's response pathway to release endorphins which produce a feeling of euphoria (Response #6, page 34).

When a feeling of well-being is obtained from a substance such as alcohol, cocaine, marijuana or other "stressors" (foods or other chemicals that the body cannot process for its benefit), the "high" lasts only as long as the body chemistry is stimulated. As the effect wears off, the person returns to his previous state and has a compensatory letdown. The body-mind psyche then craves this substance in order to regain the high.

In this manner, people become quite addicted to those foods or other substances which provide this sensation of pleasure when eaten. **This represents the second major Dietary Distress Saboteur.**

If a food that triggers this pleasurable mechanism is omitted from the dieter's weight loss regimen, it will be only a matter of time before the dieter begins to "crave" that food. In most cases it is exactly this same food that created the dieter's weight problems in the first place. These cravings will always sabotage weight loss goals.

The Biofeedback Reflex Test of Chapters Four and Five allows dieters to recognize their Unsuspected Stress Foods prior to eating them. The anti-addiction technique of Chapters Six and Seven provides the necessary information to allow dieters to eliminate their "craving" sensations for these foods.

CHAPTER FOUR

USING BIOFEEDBACK
TO
ELIMINATE SABOTEURS

THE FOOD INSTINCT REFLEX RESPONSE

Your Biofeedback Reflex
Recognizing Your Food Instinct Reflex
The Food Instinct Monitor
Monitoring and Observing Your Food Instinct Reflex
Observing Your Muscle's Reflex Reaction
The Biofeedback Reflex Test (B.R.T.)
Adventures in Personal Power

In this chapter you will learn a fascinating biofeedback technique that will allow you to recognize your Food Instinct Signal. Using this simple biofeedback technique, you will be able to observe how your body is responding to various foods. This chapter presents the instructions for the Biofeedback Reflex Test (B.R.T.) used for that process.

Notes:

Thoughts:

Actions:

Questions:

Additional References:

Collected Research Papers of Dr. George Goodheart
Touch For Health by Dr. John Thie
Behavioral Kinesiology by Dr. John Diamond
Applied Kinesiology Synopsis by Dr. David Walther

CHAPTER FOUR

Using Biofeedback to Eliminate Weight Loss Saboteurs

The Food Instinct Reflex Response

Your Biofeedback Reflex

The chewing of food does more than just chop it into smaller pieces. While mixing with saliva, the food begins the sophisticated mechanism of digestion -- of being processed into simpler components.

Think about this for a moment:

"I am holding this large juicy lemon that I just picked from the tree. Now I am cutting it in half and the juices are dripping all over my hands. I have now placed the lemon in my mouth. I am squeezing the juices onto my tongue. I am now sucking the sour juices from the lemon."

What did your mouth do while reading the above paragraph? Chances are you salivated just as if you had actually put the lemon into your mouth. Your body was responding because your central nervous system proceeded with the necessary reactions needed to digest a lemon. All of this required fine control and coordination of these reactions.

The chemical reactions that take place in your mouth initiate the many body responses needed to process your food.

If there is a problem with a food's relationship to the digestive mechanisms, a reflex signal is sent to and from the brain (central nervous system) with the message that a problem exists.

The chemical reactions that take place in your mouth (saliva and food) initiate the many necessary body responses needed to process your food. Before you swallow the food that you are chewing, many of your digestive organs are already proceeding with their own chemical reactions in order to receive that particular food. Experiments have shown that food deposited directly into a test animal's stomach without first interacting with the mouth's enzymes, just stays in the stomach and is never digested.

If there is a problem with a food's relationship to the digestive mechanisms, a reflex signal is sent to and from the brain (central nervous system) with the message that a problem exists.

It is this signaling reflex that can be observed with biofeedback procedures. By monitoring this reflex, it is possible to monitor your body's reaction to specific foods.

The Biofeedback Reflex Test, described on pages 73 - 75 allows you to observe this reflex mechanism at work.

Recognizing Your "Food Instinct" Reflex Response to "Dietary Distress"

Until the advent of scientific technology and instrumentation that allowed personal observation of body responses, you could only sense or feel your body respond to different foods. There was no visibly tangible way of detecting any of those triggers which were eliciting unsuspected subconscious responses within your body.

Now biofeedback techniques exist that demonstrate to you when this reflex has been triggered. Science has shown that you can regulate "involuntary" physical functions which can be monitored by instrumentation and then observed by you through any one of your five senses. Many involuntary nervous system functions can be brought under conscious control if you can obtain conscious information about the process.

Some of the biofeedback methods already being used to observe body responses to specific stimuli include monitoring pulse rate, temperature change, electromyograph measurement and instantaneous changes in the physical strength of body muscles. Research has proven that the muscles of your body respond instantaneously to these biological reflex signals. This effect can be observed by monitoring changes in muscular strength in response to specific stimuli.

The importance of the body musculature in relationship to the mind-body dynamics has become firmly established with the advent of a number of technologies including Bio-Energenics by Alexander Lowen, Functional Integration by Moshe Feldenkrais, the Alexander Technique by F.M. Alexander, Rolfing by Ida Rolf, Applied Kinesiology by Dr. George Goodheart and Touch For Health by Dr. John F. Thie.

The Biofeedback Reflex Test (B.R.T.) is the biofeedback method that allows you to observe this response mechanism at work.

The first instrumentation used to demonstrate this reflex (an increase or decrease in muscular strength) was the Jamar® Dynamometer (Figure 4.1). This tool is primarily used by a physician to perform neurological and orthopedic examination tests. It is a very sophisticated grip meter equipped with a gauge that allows the physician the opportunity to record in pounds and kilograms the amount of pressure exerted by the grip of a patient.

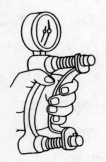

Figure 4.1
Jamar®
Dynamometer

In 1978, Dr. Alan Scopp presented a research paper to the National Center for the Transitional Healing Arts. In this study, Dr. Scopp measured and recorded the variations in muscular strength in response to environmental stimuli. Using the Jamar® Dynamometer, he was able to demonstrate the physiological reflex of muscular strength in response to environmental stimuli and to record changes in strength in response to various environmental stimuli.

With the Bio-My Kinesiometer (Figure 4.2) physicians were able to isolate and evaluate the effect of different stimuli on a single muscle as opposed to the many necessary to effect a grip. This device further allowed the physician to monitor the reflex reaction of all muscle groups to specific stimuli.

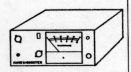

Figure 4.2
Bio-My
Kinesiometer

The Kin-Com (Figure 4.3) from Chattecx Corporation is a hydraulically powered, computer controlled exercise, testing and research system. It is used to record any induced reflex response and to produce a chart graph of the results — recording the pre and post muscle strength after the introduction of a food stimulus. Figure 4.4 shows the results of testing a muscle's ability to be influenced by various food substances introduced into the mouth.

The first trial represents the normal strength of the muscle. The second trial represents the strength measurement of the same muscle being decreased by a negative stimulus reflex. The third trial, obtained when the muscle was monitored after the stimulus was removed, shows that the muscle strength returned to its original level.

Figure 4.4 shows the results of testing a muscle's ability to be influenced by various food substances introduced into the mouth.

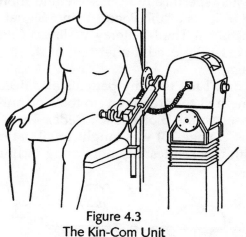

Figure 4.3
The Kin-Com Unit

Strength of Tested Muscle
0.....10..... 20....

Trial 1
Trial 2
Trial 3

Figure 4.4
The Kin-Com Graph

Figure 4.5
Preston Pinch Gauge

Because the "Food Instinct" has an instantaneous effect on all response pathway muscles, it becomes immediately observable if muscles are weakened in response to stress producing foods.

The small but sophisticated Preston Pinch Gauge (Figure 4.5) monitors any shift in the strength of the pinching ability of finger muscles. The measurement is in pounds and kilograms.

The Food Instinct Monitor
(Please refer to p. 177 - Addendum #1)

The Jamar Dynamometer was the first tool that observed and recorded instantaneous reflex changes in physical strength caused as a physical response to specific substances. The Food Instinct Monitor (Figure 4.6) was the first dynamometer created specifically to observe and monitor the reflexes of the Food Instinct Signal. Patients at The Monterey Wellness Center successfully used it after two or three attempts.

But, I felt that my patients also needed a method that required no tools or purchases so the Biofeedback Reflex Test (B.R.T. page 73) was taught to all patients. With this self-testing monitoring technique

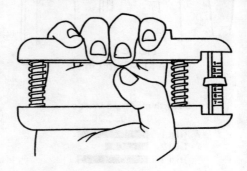

Figure 4.6
The Food Instinct Monitor

patients can team up with a partner and monitor their Food Instinct anytime they wish, free of instrumentation.

Monitoring and Observing Your Food Instinct Reflex

Because the Food Instinct has an instantaneous effect on all response pathway muscles, it becomes immediately observable if muscles are weakened in response to stress producing foods.

Although each person's response to the same food may be different, most people are quick to observe the results of the Biofeedback Reflex Test: that specific foods consistently decrease their muscle strength.

Those foods that produce a muscle-weakening response are your stress foods.

And, because they cannot be digested well, they will sabotage your weight loss goals.

By using the Biofeedback Reflex Test (B.R.T.) you can train yourself to become consciously aware of your automatic subconscious "intuition" concerning the foods in your diet.

It is the awakening of your awareness of this reflex signal that is important.

It is this newly gained awareness of your Food Instinct that gives you the freedom to choose between a weakened or empowered condition. It is the freedom to choosebetween failure or success in completing and maintaining the positive results of a weight loss program.

By using the Biofeedback Reflex Test (B.R.T.) you can train yourself to become consciously aware of your automatic subconscious "intuition" concerning the foods in your diet.

71

It is this newly gained awareness of the "Food Instinct" that will give you the freedom to choose between a weakened or empowered condition.

Observing Your Muscle's Reflex Reaction to Your Personal Food Stressors

The Biofeedback Reflex Test (B.R.T. Figures 4.7 and 4.8) allows you to observe a fascinating and complex biological reaction. The simplicity of the biofeedback technique used to monitor this reaction may give some readers a false sense of ability to use and interpret the results. Always remember that this is a delicate, neurological biofeedback signal. Your willingness to practice and master the B.R.T. will determine the effectiveness of this great tool.

Using the B.R.T., you can observe any decrease in your muscle strength in response to foods. It is this observation which will allow you to recognize those foods that will enter your system as complex molecules. Recognizing these foods prior to eating them gives you the much needed key to prevent Unsuspected Stress Foods from sabotaging your weight loss goals.

Your goal is to discover those foods that influence your B.R.T. You will do this by observing which foods cause a decrease in your B.R.T. Eventually you will develop a deep awareness within you that will allow you to recognize when a weakening has taken place.

The Biofeedback Reflex Test (B.R.T.)

Step 1
Remember that you will be monitoring a very delicate neurological reflex by observing a biofeedback mechanism that effects your muscles. Plan to monitor this reflex during a quiet time with a minimum of distractions.

Step 2
Stand quietly with your friend who is doing the testing by your side (Figures 4.7 and 4.8). It does not matter if he stands to the right or left side.

Step 3
Stand erect with your arms by your side. Place the thumb of the hand that you will be monitoring next to your leg. If you are wearing pants, place your thumb along the outside seam of those pants.

Step 4
Have your friend who is standing by your side stabilize you by putting one hand on your shoulder.

Step 5
Have your friend place two or three fingers of his other hand along the inside of your testing arm above your wrist (between your arm and leg).

Step 6
Using a significant amount of strength, press your arm against the side of your body to squeeze your friend's fingers. Concentrate and focus as much strength as possible to maintain this position.

The Biofeedback Reflex Test (B.R.T.) allows you to observe a fascinatingly complex biological reaction.

It is a decrease in your B.R.T. that allows you to recognize those foods that enter your system as complex molecules.

Step 7
Have your friend, whose fingers are between your arm and leg, exert a MINIMAL amount of strength as if trying to pull your arm away from your body.

Your friend should not actually pull your hand away from your body.

Your goal is to to recognize the amount of resistive strength you can offer to your friend's gentle pulling against your arm.

Practice this to become aware of your RESISTIVE CAPABILITY TO THE GENTLE PULL.

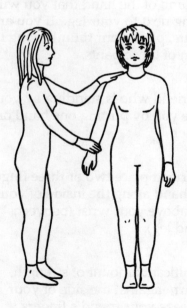

Figure 4.7
The Biofeedback Reflex Test (Using Right Hand)

Remember that you are monitoring your own awareness of any decreased resistive capability.

Proceed with the food monitoring process explained on page 81.

After some practice in using this bio-feedback technique, you will be able to discover your Unsuspected Stress Foods -- those foods that make you weak, rob you of your strength or take you out of control. Soon you will be identifying those stress-producing foods capable of sabotaging your weight loss goals.

The recognition of Unsuspected Stress Foods prior to eating them gives you the necessary key to prevent these foods from sabotaging your weight loss goals.

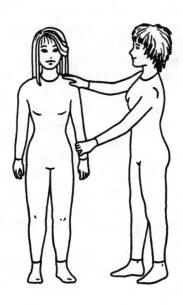

Figure 4.8
The Biofeedback Reflex Test (Using Left Hand)

The B.R.T. points out to your conscious mind what your subconscious mind has been screaming out to you all along.

An Adventure in Personal Power

This biofeedback technique monitors your body's response to food and is based on the principles of Applied Kinesiology. You will be observing a decrease in your muscular strength that occurs when you introduce a stress-producing food into your system.

As with any very complicated and complex scientific observation that has been overly simplified for the general public, there are exceptions to the rule. If you want further information or interpretations of your results, you can contact any doctor and/or clinic that uses the art and science of Applied Kinesiology as part of its health regimen. You can find this information in the directory at the end of this book (page 195).

You will, like most people, require a few days to become comfortable in trusting the biofeedback signals you will be personally observing. Like any new tool, the more you use it, the better you will become at interpreting the results. Alerting yourself to your own Food Instinct Signal is well worth the short trial and error period necessary to develop this skill. Once you witness the consistency of your responses, you will begin a fascinating adventure in personal power.

Remember, the B.R.T. points out to your conscious mind what your subconscious mind has been screaming out to you all along. It is not uncommon for a person to say, "I was suspicious of that food." or "My intuition has been telling me that."

CHAPTER FIVE

ELIMINATING YOUR UNSUSPECTED FAT PRODUCING STRESS FOODS

THE BIOFEEDBACK REFLEX TEST (B.R.T.)

Recognizing Dietary Distress
B.R.T. Instructions
Developing the Skill to Monitor Your Food Instinct
The Menu Monitoring Procedure

By the time you start this chapter you will have practiced and mastered the basic Biofeedback Reflex Test. It is now time to use this biofeedback technique to monitor the foods in your diet. This chapter gives you the directions you need to monitor your foods and provides a sample demonstration of how to monitor breakfast, lunch and dinner.

Keeping Track of Your Weight Loss Saboteurs

Week #1 - Unsuspected Stress Foods

Week #2 - Unsuspected Stress Foods

Week #3 - Unsuspected Stress Foods

Week #4 - Unsuspected Stress Foods

Additional Reference:

Touch for Health by Dr. John Thie
Applied Kinesiology Synopsis by Dr. David Walther
Your Body Never Lies by Dr. John Diamond
Collected Research Papers by Dr. George Goodheart
Weight Watchers Food Plan Diet Cookbook by Jean Nidetch

CHAPTER FIVE

Eliminating Your Unsuspected Fat Producing Stress Foods

The Biofeedback Reflex Test (B.R.T.)

"Dietary Distress" is a causative factor that can initiate physical stress, anxiety and depression. The worst aspect about this type of stressor is that the victim does not recognize the fact that the food he considers "good" is causing problems for him. Therefore, he places the blame for his "daily stress" on the wrong aspects of his life."

Dr. Joel S. Shain
Life Empowerment

Recognizing Dietary Distress

If you could recognize specific foods that would sabotage your weight loss goals before you even swallowed them, your dieting and weight loss worries would be simplified. Earlier chapters demonstrated that the central nervous system is equipped to receive and interpret biofeedback signals from digestive organs in response to a food you are eating.

Now you will learn to use the Biofeedback Reflex Test (B.R.T.) to provide visible evidence when a specific food will not be handled properly by your digestive system. The B.R.T., by showing immediate muscle response to your Food Instinct Signal, teaches you to recognize your specific "stressors" (i.e., "weight loss saboteurs") before you even swallow them.

Recognize specific foods that will sabotage your weight loss goals before you even swallow them.

The B.R.T. provides visible evidence that a specific food will not be handled properly by your digestive system.

The B.R.T. gives the dieter an additional tool for successful weight loss, one that has been missing from all diet programs.

At this point I am sure you are expecting a list of what to eat and recommendations for specific diets. You won't see such a list here. The merits of all available diets have been debated for years. Some may work well for one person, but not for someone else. Some people digest vegetables very well, others digest meat and potatoes better and some individuals can live wonderfully on rice and grains. The reason is that each person's digestive ability is based on their personal lifestyle, stress factors, heredity and illness.

Because of this, I recommend that when you choose a weight loss diet, choose one that appeals to you. Then use the B.R.T. to record which foods will be correctly digested on any given day and which will not.

Few of us have the ideal situation of being so finely attuned to our body dynamics that we can "hear" our own biofeedback signaling.

But this is a skill we can learn and until we do, we are fortunate to have the B.R.T. to recognize which foods we need to avoid.

So, if you are following a specific diet, (Weight Watchers, Nutri-Systems, Low Carbohydrate, Air Force, Fit for Life, Athletic, etc.) remember that each of these programs is valuable and effective for weight loss. But, by using the B.R.T. on a daily basis to monitor your body's response to each food in your diet program, you will discover those foods that would sabotage your weight loss goals on any specific day.

The B.R.T. Instructions

Developing the Skill
Monitoring Your Food Instinct Signal

Step 1
Place a maximum of ten food items in front of you. If an item contains a combination of foods, test the combined food as a single item.

Step 2
Complete steps #1 thru #7 of the B.R.T. (page 73) with no food in your mouth to determine your baseline strength.

Step 3
Take a small bite of the first food. Chew it a few times and hold it under your tongue. With the food still under your tongue, again, complete the B.R.T. Note if there is any deviation from your original strength with no food in your mouth. Is there any change in your ability to maintain your strength? Is it harder to do it? Is it easier? Did that food make your muscle response weaker? Did it have any effect at all?

Step 4
Thoroughly rinse your mouth with water.

Step 5
Now with your mouth empty again, test to reconfirm your original baseline strength.

Step 6
Test your muscle response to food #2. When finished, rinse again. Continue this sequence until you observe how you are affected by each food.

B.R.T. questions: Did that food make your muscle response stronger or weaker? Did it have any effect at all?

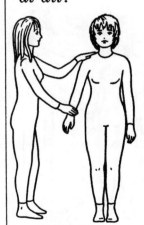

Figure 5.1
Biofeedback
Reflex Test

Observe the increase or decrease in your B.R.T. strength caused by your weight loss saboteurs.

Step 7

Keep a record of all foods that consistently weaken your muscle response. See if you can determine why these foods are affecting you. You may want to check with a nutritionist to see if these foods fall into any specific category.

Step 8

Avoid those foods that decrease your strength. These foods are a source of unnecessary stress to your system. They are the foods that will sabotage even the best weight loss and weight maintenance programs.

Again, to reiterate, using the B.R.T. to monitor your body's response to food is based on the principle of Applied Kinesiology. The observation that you can make is a decrease in muscular strength caused by the introduction of specific food into your system. As with any very complicated and complex scientific observation that has been simplified for the general public, there are exceptions to the rule. Individuals who want further information or interpretations of their results are invited to contact any doctor and/or any clinic that uses the art and science of Applied Kinesiology as part of its health regimen (Please refer to the support personnel directory on page 194).

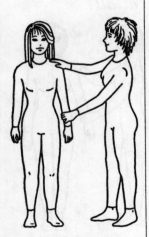

Figure 5.2
Biofeedback
Reflex Test

The Menu Monitoring Procedure

The following meals are given as an example of how to monitor your Food Instinct. These sample meal menus were taken from a Weight Watchers plan and individual meals were selected at random from different days of the menu.

Breakfast
Chilled Tomato Juice, 1 cup
Soft-Cooked Egg, 1 large
Pumpernickel Toast, 1 slice
Reduced-Calorie Margarine, 1 tsp.
Skim Milk, 1 cup
Coffee or Tea

Prepare breakfast as itemized and place all foods on the table along with a large glass of water. Establish the baseline strength of your B.R.T. Take a small sip of the tomato juice and roll it around in your mouth. Wait five seconds. While holding it under your tongue, monitor your B.R.T. Note your response. Rinse your mouth with water and reconfirm your ability to re-achieve the original strength.

Take a small bite of the cooked egg and chew it a few times. Hold it under your tongue. Complete the B.R.T. Note your response. Rinse your mouth and reconfirm your original baseline strength.

Next take a small bite of the toast and chew. Hold the food under your tongue as you complete the B.R.T. Note your response. Rinse your mouth well and recheck your baseline. Repeat the procedure with the margarine.

The search for your "Food Instinct" is a true adventure in personal power.

The B.R.T. awakens your awareness of your biofeedback alarms in response to weight loss saboteurs.

Take a sip of the milk and roll it around in your mouth. Hold it under your tongue as you test your response. Rinse your mouth and recheck baseline. Repeat for coffee or tea.

Keep a written record (page 78) of those items which reduced your muscle-strength reading and do not eat those foods until a subsequent testing shows them to be neutral.

Lunch
Grilled American Cheese Sandwich
Sliced Cucumber and Tomato
Cole Slaw, 1/2 cup
Green Grapes, 12 large
Diet Cherry Cola

Prepare for lunch by placing all of the foods on your table along with a large glass of water. Determine the baseline strength of your B.R.T.

Take a small bite of the grilled cheese sandwich and chew it a few times, holding it under your tongue as you monitor your B.R.T. Note your response. Rinse your mouth as instructed and reconfirm your original baseline strength. (Note that food in different forms, i.e., cooked vs. uncooked, will affect your body differently. Do not assume that a clear reading for raw cheese will hold true for grilled cheese.)

Take a small bite of the cucumber and chew it a few times. Hold it under your tongue as you monitor your results. Note your response. Rinse your mouth and re-check your baseline strength. Repeat the procedure with the tomato.

Next take a small bite of the cole slaw. Hold it under your tongue as you monitor your B.R.T. Note your response. Rinse mouth and recheck your baseline strength.

A cautionary note: Be aware of your arm muscle becoming fatigued during testing, for muscle fatigue creates inaccurate results. Should you notice arm fatigue, the solution is simple. Continue the test using your other arm.

Taste some grapes. Hold them under your tongue as you monitor your response. Note your response. Rinse your mouth and recheck your baseline strength.

Take a sip of the diet soda and roll it around in your mouth. Hold the soda under your tongue as you monitor your B.R.T. Rinse your mouth. Recheck baseline.

Keep a written record (page 78) of those items which reduced your muscle-strength and do not choose those foods until a subsequent testing clears them.

Become aware of the internal feelings that you experience when you monitor a negative food.

Dinner
Mixed Green Salad
1/2 Tbs. Blue Cheese Dressing
Broiled Hamburger, 4 oz.
Ketchup, 2 tsp.
Mashed Potato, 3oz. with
1 Tbs. sour cream and snipped chives
Steamed Spinach
Pineapple Chunks, 1/2 cup
Coffee/tea

Before mixing the salad, place all of the ingredients on the table along with a large

Foods that are sabotaging your weight loss goals should be eliminated from your diet.

glass of water. Establish your baseline muscle strength of your B.R.T. Take a small bite of the lettuce and chew it a few times. Hold it under your tongue as you monitor your results. Note your response. Rinse your mouth as instructed and recheck your baseline strength. Repeat with a small bite of the other ingredients. Test your dressing separately.

Prepare all of your foods of your dinner menu. Take a small bite of the cooked hamburger patty and chew it a few times. Hold it under your tongue as you monitor your B.R.T. Note your response. Rinse your mouth and recheck your baseline strength. Repeat with the ketchup.

Is your arm muscle fatigued? If it is, do not proceed to test the other items with the same arm or the testing may not be accurate. Use your other arm for the remainder of your food testing.

Take a small bite of the mashed potato. Hold it under your tongue as you monitor your B.R.T. Note your response. Rinse your mouth and recheck your baseline strength. Repeat with the the sour cream and then the chives separately.

Take a taste of the spinach. Hold it under your tongue as you complete the B.R.T. Note your response. Rinse your mouth and recheck your baseline strength. Monitor your response to the pineapple chunks and your beverage.

Keep a written record (page 78) of those items which reduced your muscle-strength reading and do not choose those foods until a subsequent testing shows them to be neutral.

CHAPTER SIX

ADDICTIONS
&
THE STRESS MECHANISMS

ADDICTED TO THE STRESSES OF WEIGHT LOSS

Life Empowerment
Addicted to Foods, Overeating and Hunger
The Addiction Erasure Procedure
Addicted to the Response Mechanism
Replacing Addictions with Balance and Peace
The Seven Anti-Addiction Activities
The 60-Second Anti-Addiction Procedure

Chapter Six explains the 60-second anti-addiction procedure used to erase "cravings" for foods, thoughts or actions created through the addiction process. This technique calms, soothes and subdues the addictive desire. This chapter gives the preleminary explanation of each of the seven parts of the procedure. Then, Chapter Seven explains in detail how to complete each of the seven parts of the anti-addiction activities.

Notes:

Thoughts:

Actions:

Questions:

Additional References:

Joy's Way by Dr. Brugh W. Joy
Touch For Health by Dr. John F. Thie
Life Empowerment by Dr. Joel S. Shain
Behavioral Kinesiology by Dr. John Diamond
The Theory of Applied Electro-Acupuncture by Dr. Thomas Wing
Breakthrough for Dyslexia & Learning Disabilities by Dr. Carl Ferreri

CHAPTER SIX

Addictions
&
The Stress Mechanisms

Addicted to the Stresses of Weight Loss

"If I wish to release a handful of sand all I need do is open my hand and the sand takes care of itself. ... to heal myself all I need do is stop the abuse."

Michael Blate
The Tao of Health

This 60-second technique enables you to control and eventually erase your food addictions.

Life Empowerment

Chapters Five and Six detail a 60-second technique which enables you to control and eventually erase your addictions - whether they are to foods, overeating or even the "sensation" of hunger. The activities of the addiction erasure procedure are based upon the Empowerment Patterning Technique. I developed this technique (explained in my book, *Life Empowerment)* to give my patients a tool by which they could return the systems of their Response Pathway Mechanisms (page 44) back to a non-addictive condition.

By using this technique, my patients were able to erase, for a period of time, any feelings of cravings resulting from addictions to drugs, alcohol, thoughts, actions and even foods. The "craving" desires are replaced with sensations (feelings) of balance, peace,

The addiction erasure procedure is like an aspirin for those addictions triggered by stress, anxiety, and tension. It provides a few hours of relief from the craving desires.

and energy. It is the normalization of physical electromedical readings (Figure 6.1 and 6.2) that creates this energized, balanced, peaceful state of being.

Addicted to:
- Foods -
- Overeating -
- The "Sensation" of Hunger -

As you will learn in Chapter Seven, an addiction is a neurologically ingrained, conditioned reflex response to certain foods, thoughts or events. Addictions to foods - whether it be in the form of excess (overeating), or cravings (for specific or multiple foods) are often rooted in earlier (guilt-producing) events/programming of your life. The event could have happened only once or had many repetitions. At that time in your life, for reasons based on your lifestyle, stress factors, heredity or illness, this event/events set up a pathway in your nervous system to allow you to deal successfully with that event (i.e., The Slap - page 41). It was this pathway that gave you the correct "feeling", "sensation", "intuition" or "sense" to allow you to respond correctly.

Most of us are not aware of the many addictions we carry around in our subconscious memory. These reaction patterns can affect us on either our emotional, spiritual or physical plane. In many ways our adult life is functioning still within that child's vision that is locked deep within our subconscious mind.

The Addiction Erasure Procedure

The addiction erasure procedure is like an aspirin for those addictions triggered by stress, anxiety, tension and other emotional states. It provides a few hours of relief from the "craving" of your food addictions.

It gives you a tool to suppress a major weight loss saboteur when it starts signaling negative behavior.

If repeated over a thirty-day period -- each time there is a craving for addictive foods -- the intensity of that craving's "desire signal" will be significantly or permanently reduced.

When you are feeling a "craving" for the foods that you have become addicted to, rest assured that the Conditioned Response Pathway (Figure 2.5, page 44) is being fired. The activities of the anti-addiction program are designed to erase unnecessary conditioned behavioral patterns by replacing the triggering of the response pathway with a sensation of peace, balance and harmony.

Addicted to the Response Mechanism

As we observed in the earlier example of the television contestant (page 32), the triggering of the Response Pathway Mechanism is meant to be very pleasurable. When we must choose between survival or extinction, the automatic decision to survive is sweetened by the rewards of the mechanism. An added benefit of the joy of winning is the thrill of survival.

When you are feeling a craving for a food not included in your diet program, this life empowering activity gives you a tool to suppress that urge.

The excitement of a new job, a new relationship, or a new change of any kind in your life triggers your Response Pathway Mechanism.

The body experiences and enjoys the firing of the Response Pathway Mechanism as a high and can actually crave its repetition.

The excitement of a new job, a new relationship, a new food or a new change of any kind in your life triggers this mechanism until you, as a physical organism, are assured of your safety and comfort with the change. Until you achieve balance and harmony with any life change, the Response Pathway Mechanism is continually activated when confronted with that change.

Depending on how long it takes you to achieve this balance, you may become addicted to the stimulation of your response mechanisms.

If your body cannot develop harmony and peace with a specific trigger (please refer to page 62 - The Second Major Dietary Distress Saboteur), it continually stimulates the pleasurable survival mechanism every time you confront this trigger. Whether the trigger is a food or substances such as drugs, alcohol or cigarettes, you may continue to ingest these because of the rush of pleasure from the Response Pathway Mechanism - not the pleasure of the triggering agent. If you stop taking this stimulating trigger, you may crave that high and seek it out again.

You are, in actuality, craving your body's response to the trigger. If you continue this craving, you become addicted to your body's response to the stimulating trigger.

This addiction stage is eventually replaced with a habit phase. The physical act caused by the addiction becomes a learned, conditioned, neurologically ingrained reaction. Continual triggering may cause the response to become exhausted. The rush of excitement is no longer there, but the reflex continues automatically. At this point, if you try to eliminate a negative food from a diet regimen, you will have as much trouble as a drug addict trying to "cold turkey" his addiction.

For any diet regimen to enjoy long-term success, it is necessary to end any of this type of destructive activity that may exist.

When you are feeling a craving of "hunger" or for a food not included in your diet program or for any of your Unsuspected Stress Foods, this anti-addiction activity will give you a tool to suppress that urge.

Until you achieve balance and harmony with life changes, the Response Pathway Mechanism is continually activated.

Replacing Addictions with Balance and Peace

There are seven pathways that combine to produce a conditioned addictive response pattern. As presented on page 44, the seven responses are; the muscle response (bracing); the adrenal gland (energy) response; the respiratory/circulatory response; the thymus gland (immune system) response; the pituitary gland (psychological) response; and the central nervous system response. It is well researched and documented that when an individual is experiencing an addictive craving, these seven systems show (present) a specific pattern that can be

Depending on how long it takes you to achieve this balance, you may become addicted to the "stimulation" of your Response Mechanisms.

shown graphically on an oscilloscope screen (Figure 6.1). The wildly eratic signals of this condition contrast vividly with an individual in non-addictive condition. When attached to multiple electrodes, an individual in the non-addictive state shows an oscilloscope pattern that resembles smooth waves (6.2).

With measuring devices of this kind, doctors at The Monterey Wellness Center were able to chart body responses to various self-aided treatments as a means of sedating the triggering mechanisms of addictions. After years of research, we were able to determine specific self-treatments that had a normalizing effect on each of the seven triggered Conditioned Response Pathway systems. When we taught these self-treatment procedures to our patients they were able to control addictions by subduing the craving desire. The self-treatment erased the Conditioned Response Pathway signals triggered by their addictive desire.

As described in previous chapters, it is the normalization of the Response Pathway Mechanisms (Figure 2.2) that shuts off the Conditioned Response Pathways (Figure 2.5). This self-treatment allowed patients to exercise and develop self-control, thus reducing their addictions on their own.

This self-treatment activity replaces the craving sensation with a sensation of balance and peace. When the combined pathway mechanisms have been triggered, the following anti-addiction activities restore each of the specific response pathway systems back to their non-addictive condition.

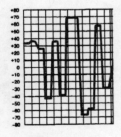

Figure 6.1
Electro-Activity
Patterns of Addictions

The Seven Anti-Addiction Activities

Activity #1
Focus your "mind's eye" on the addiction you wish to subdue. Picture yourself eating the food you are craving.
(Addicted to Negative Behavior, P. 127)

Your first task is to mentally focus on the action or addiction that you would like to subdue. By bringing up a picture of yourself engaging in this particular activity in your "mind's eye", you will be directly affecting those physical pathways that allow this conditioned behavior action pattern to take place.

Activity #2
Remove the "muscle bracing" response of your Conditioned Response Pathway. This activity removes muscle bracing that occurs in response to your addictive food.
(Response #1, P. 33 & P. 105)

The second activity consists of learning to stand in a specific position that relaxes all the body's muscles. This stance is the center of gravity position of all martial arts. When a person is standing in such a balanced position, it is virtually impossible for an attacker to move him. According to the martial arts philosophy, when the spine is firmly rooted and balanced, the bilateral "mirror image" aspects of the muscles on your right side are forced into an equally balanced condition in relation to the "mirror image" muscles on the left side. In this balanced position it is impossible for your body to retain muscle bracing.

Each of these anti-addiction activities normalizes and balances one specific Response Pathway Mechanism.

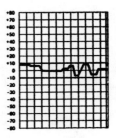

Figure 6.2
Electro-Activity Pattern
of Non-Addiction

This activity normalizes the right and left hemispheres of your brain allowing them to function better.

Activity #3
Restore your respiratory and circulatory responses (triggered by the addictive food stimulation of your Conditioned Response Pathway) to a non-addictive condition.
(Response #3 & #4, P. 33 & P. 106)

The breathing patterns for this activity are derived from the Yogic studies of energy. According to Yogic philosophy, this breathing technique enables you to breathe life into the vital areas of your heart, solar plexus and sexual areas. This breathing activity opens and energizes your response mechanisms. There are six breathing cycles. The first breath expands your chest. The second breath expands your stomach. The third breath expands your lower abdomen. For the last three breaths, the process is repeated in reverse order.

Activity #4
Free the thought pattern interchanges between the right and left hemispheres of your brain that are responsible for locking you into a programmed response to an addictive behavior pattern.
(Response #8, P. 35 & P. 107)

This fourth activity is the result of research by Doman and Delacato on dyslexia. The researchers observed and reported the importance of the crawling stage in child development. The right and left side of the brain must work together for adequate brain functioning. The Cross-

Crawl of Activity #4 stimulates those areas in the brain where fibers from the right brain cross over and meet those of the left brain, allowing free movement between the two hemispheres.

In practical experience, if an executive gets locked into his analytical left brain and cannot get into his creative side, he may drink an alcoholic beverage to help shift. The same is true of the artist who has used his creative mind all day. He may use drugs to shift into his linear side so that he can tend to his business affairs. This activity accomplishes the same effect.

Activity #5

Use acupressure to restore the addictive conditioned reflex pathway of your adrenal glands back to a non-addictive condition. (Response #2, P. 33 & P. 109)

For the fifth activity, I acknowledge Dr. Brugh Joy, M.D. for developing the original method used to rebalance triggered adrenal gland pathway responses. With slight modification, a simple tapping technique of nerve endings and acupressure points has the net effect of soothing the adrenal gland's electro-medical response pathways. This activity gives the adrenal glands a better opportunity to cope with stress factors such as exhaustion and burn-out.

This self-treatment allows your adrenal glands to successfully handle stress factors such as exhaustion and burn-out.

97

Thie's studies proved it to be beneficial in the relief of fears, nightmares, frustrations and other problems of emotional and mental strain.

Activity #6
Use acupressure to restore the addictive conditioned reflex pathway of your thymus gland back to a non-addictive condition.
(Response #5, P. 34 & P. 109)

In his book *Behavioral Kinesiology*, Dr. John Diamond, M.D. describes methods of stimulating the thymus gland which help strengthen and rejuvenate the immune system. Given this treatment, his patients overcame stress-related problems more quickly. His classic "thymus thump" is used as the basis for the sixth of our addiction-erasing activites.

Activity #7
Calm and restore the emotional and psychological responses of the addictive reflex pathway of your pituitary gland back to a non-addictive condition.
(Response #6, P. 35 & P. 110)

The technique of using light pressure on neuro-vascular holding points on the forehead to release anxiety, stress and tension comes from Dr. John F. Thie's book, *Touch For Health*. These holding points are acupressure points related to the emotional centers. This treatment tends to neutralize, sedate and calm emotional upset. Thie's studies proved it to be beneficial in the relief of fears, nightmares, frustrations and other problems of emotional and mental strain.

The 60-Second Anti-Addiction Patterning

You're sitting at home watching the T.V. It has been a few hours since you had lunch. Dinner is not scheduled for two more hours. Your mind starts envisioning that extra burrito that the kids didn't eat last night. The same burrito (470 calories) that you've made for years and always enjoy eating. Your urge to go to the refrigerator so that you could take a few bites has forced you to stand up and you find yourself walking to the refrigerator. This is the time for the anti-addiction procedure. Your Conditioned Response Pathway Mechanism is triggering those necessary mental pictures that will cause you to re-create an unnecessary and unneeded behavioral action. It is while this mind-body dynamic is in full action, that the anti-addiction patterning technique will calm, soothe and subdue the conditioned reaction pattern that is about to force you to eat that burrito.

Stand quietly for a moment and picture in your "mind's eye" the action (eating) that you are about to perform. Now, gently bend and adjust your knee position to redistribute your weight. Then comfortably breathe in. The first in-breath expands your chest, the second breath expands your stomach, and the third expands your lower abdomen. The breathing process is then reversed for the last three breaths.

Then for 30 seconds, march in place, raising opposite arms and legs as you turn

Your Conditioned Response Pathway Mechanism is triggering those necessary mental pictures that will cause you to re-create an unnecessary and unneeded behavioral action.

In this 60-second exercise you normalized the electro-medical activity patterns of your Response Pathway Mechanisms.

your head toward the raised arm. Swing your limbs high as you march in one spot for about 20 steps.

Now thump the finger pads of your right hand in a counter-clockwise circle around your navel. Make ten circles which transverse the circumference of your belly, crossing the lower ribs and lower abdomen.

In Tarzan fashion, thump your chest rapidly with the finger pads of your right hand. Do this slowly and gently.

Then bring both hands up to your forehead and very gently rest your finger pads on your forehead for 20 seconds.

Keeping your knees slightly bent and retaining your relaxed mood, gently redistribute your weight evenly on both feet.

This completes the exercise. You will feel wonderfully relaxed and refreshed. In this 60-second exercise, you normalized the electro-medical activity patterns of your Response Pathway Mechanisms. You created a state of balance within your mind-body dynamics.

This process comprises the anti-addiction treatment. Chapter Six gives a detailed explanation of how to perform each activity.

ELIMINATING FOOD ADDICTIONS NOW!

RESTORING YOUR RESPONSE PATHWAY MECHANISMS

Chapter Seven continues the food addiction discussion by teaching you how to do the entire addiction erasure procedure. This anti-addiction technique subdues your food cravings. It should be used anytime you become aware of cravings for foods that you should not be eating. After 30 days of this kind of conditioning, most food addictions can be totally erased.

Notes:

Thoughts:

Actions:

Questions:

Additional References:

Touch for Health by Dr. John F. Thie
Life Empowerment by Dr. Joel S. Shain
Addictive Behaviors by Dr. William R. Miller
Your Body Doesn't Lie by Dr. John Diamond
Between Health and Illness by Dr. Barbara B. Brown

CHAPTER SEVEN

Eliminating Food Addictions - Now!

Restoring Your Response Pathway Mechanisms

"The real issue is the integration of your external and internal universe."

Dr. Barbara B. Brown
Stress and the Art of Biofeedback

An addiction is a conditioned reflex response to a certain food, thought or event.

Do this 60-second activity any time you feel a "craving" for unnecessary or unwanted actions or foods.

Erasing Addictions

A food addiction is a neurologically ingrained, conditioned reflex response to certain foods, thoughts or events. Neurologically, a conditioned, learned, automatic, subconscious response is initiated and carried out by a specific firing of the six Response Pathway Mechanism organs. Each of these six Response Pathway Mechanism organs fire simultaneously at their own intensity setting. The combined firing produces a unique signaling to your conscious brain. It is this combined signal that presents a picture in your mind's eye of a specific desired action.

First contemplate the addictive behavioral reaction, then use the anti-addiction activity to calm, soothe and subdue the mental and emotional cravings and desires.

Each of the following activities is designed to calm, soothe, re-balance and restore to normal the addictive behavioral patterns called up through your Conditioned Response Pathway. Each activity treats one specific pathway.

This series of seven separate activities can be performed in a few minutes and can quiet and subdue cravings caused by your food addictions. Some of the activities involve simple tapping with your fingers using approximately ten pounds of pressure. You can safely "guesstimate" this pressure. To help you learn what ten pounds of pressure feels like, press your finger tips on a bathroom scale. When doing the tapping, hold your fingers together in a cone shape as in Figure 7.1.

Figure 7.1
The Cone

Activity #1
Focus your mind's eye on the addiction you wish to subdue. Picture yourself eating the food you are craving.

Mentally focus on the action or addiction that you would like to subdue. Imagine in your mind's eye a picture of yourself engaging in this particular activity. By doing so you will be directly affecting those physical pathways that allow this conditioned behavior action pattern to take place. Now picture yourself refusing to perform the action (i.e. eating the pictured food, smoking a cigarette, drinking alcohol, etc.). Now try to hold that picture while completing the following erasure activities.

Figure 7.2
Stance - Front view

Activity #2
Remove the muscle bracing response of your Conditioned Response Pathway. This activity removes muscle bracing that occurs in response to your addictive food.

First you must learn to recognize the sensation produced when standing with your body balanced in your center-of-gravity position. This is the "Hara" position common to all martial arts. To experience this position, stand with your feet slightly apart and distribute your weight equally and evenly between the heels and the balls of each foot (Figure 7.2 and 7.3)

While relaxing all the muscles of your back, bend your knees slightly, keeping your spine straight and vertical. The pelvis should be tucked under so that the buttocks is not throwing your upper body forward (7.3).

Relax your chest, shoulders, arms, hands, face and back. Stop for a moment and concentrate on relaxing these muscles. Breathe slowly and evenly.

Imagine the whole weight of your body sinking down (although your spine remains straight) to an area in your lower lumbar spine (just above your hips). Feel the firm structure of your feet, legs, and hips supporting your relaxed and balanced upper body.

Move a little from side to side, forward and back, twisting and gently shaking yourself until you feel "centered" or focused at that point on your spine above your hips. Concentrate your attention at this point. Keep relaxed. Continue to breathe as you recognize this area in your lower spine as your center of gravity.

Mentally focus on the action or addiction that you would like to subdue.

Figure 7.3
Stance - Side view

Breathe in love; breathe in forgiveness; breathe in whatever you need.

Imagine that this area of your body is the source of all your natural energies. Imagine a warm force spreading from here throughout your entire body, slowly reaching up to your scalp and down to your toes. Try to actually "feel" it! Many people feel this warm spreading force after a few days.

Activity #3
Restore your respiratory and circulatory responses (caused by the addictive food stimulation of your Conditioned Response Pathway) to a non-addictive condition.

Each of the six breaths of this activity has three components: the in-breath, the retention, and the out-breath. Be sure to remain standing in your center-of-gravity position while doing this and other activities.
Inhalation - During inhalation, imagine yourself absorbing life-giving energies. Breathe in love, forgiveness, joy, acceptance, enthusiasm or creativity. Breathe in whatever you need.
Retention - Hold the retained air for a count of five seconds. During retention, focus on extracting the gift of life from the air and circulating it throughout your entire body.
Exhalation - Exhale slowly. During exhalation, imagine all negativity and tension leaving your body and returning to the earth for purification. Breathe out exhaustion; breathe out hatred; breathe out whatever you no longer need.

Do this breathing sequence slowly. There are six breaths involved. Maintain your stance in your center-of-gravity position. Visualize the first breath being drawn into your chest area; visualize the second being drawn into your stomach area; visualize the third being drawn into your abdominal area.

Reverse this order for the fourth, fifth and sixth breaths; visualize the fourth breath filling your lower abdomen; visualize the fifth breath filling your stomach area; and, visualize the sixth breath filling your chest area.

Breathe out exhaustion; breathe out hatred; breathe out whatever you no longer need.

Activity #4
Free the thought pattern interchanges between the right and left hemispheres of your brain that are responsible for locking you into a programmed response to an addictive behavior pattern.

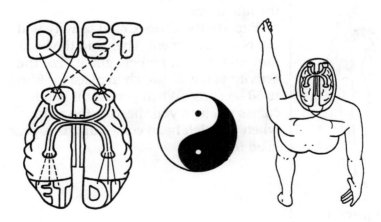

Figure 7.4
The Cross-Crawl and the Hemispheres of the Brain

107

Figure 7.5
The Cross-Crawl
left hand - right leg

Figure 7.6
The Cross-Crawl
right hand - left leg

A wonderful body movement called the Cross-Crawl enables the linear logical, arithmetic and speech functions of your brain's left hemisphere to merge with the pictorial, intuitive, artistic, and tactile functions of the brain's right hemisphere.

Fifteen to 30 seconds of Cross-Crawl activity will help return the balance between hemispheres. It enables you to use your entire brain while you alternately stimulate and relax your muscles.

The Cross-Crawl is modeled after the correct crawling motions of an infant with some therapeutic modifications. Normal crawling or creeping involves moving the right arm and the left leg forward at the same time, and then bringing the left arm and the right leg forward at the same time. In the standing position this looks like marching in place with one slight addition; you will be turning your head to the side of the raised arm.

First try the Cross-Crawl activity without head movement. Then add the head movements after you get going. For this activity you will march in place for fifteen to 30 seconds. When your right hand comes up, turn your head to the right; when your left hand comes up, turn your head to the left.

Activity # 5
Use acupressure to restore the addictive conditioned reflex pathway of your adrenal glands to a non-addictive condition.

The adrenal glands are very important in helping your body successfully cope with the stressors in your life. To normalize your adrenal gland pathway you will be tapping on the acupressure pathway that surrounds the adrenal area. Use either hand and apply about ten pounds of pressure as you tap in a circular pattern starting at the belly button. Move up the left side of your body tapping up to the area just under the lower rib cage. Then tap down your right side, still following the bottom of the rib cage. Continue tapping back to your starting position at the belly button.

You are using this circular tapping to stimulate the acupressure points of the adrenal glands. Continue tapping this area for up to 30 seconds or 100 taps.

Figure 7.7
The Adrenal Balance

Activity #6
Use acupressure to restore the addictive conditioned reflex pathway of your thymus gland to a non-addictive condition.

The thymus gland is located in an area under your breast bone in the upper part of your chest. It can be located by searching for a slight bump about one or two inches down from the top of your breast bone.

Use the tips of your fingers to tap the acupressure points in this area which are

Figure 7.8
The Thymus Balance

**Figure 7.9
The Pituitary Balance**

directly over that bump. Use ten pounds of pressure for this treatment and stimulate the area for up to 30 seconds or 100 taps.

Activity #7
Calm and restore the emotional and psychological responses of the addictive reflex pathway of your pituitary gland to a non-addictive condition.

Place the fingers of your right hand on the right side of your forehead between your hairline and your eyebrow. Search for a little bump (called the frontal eminence) that exists on your forehead above your eye and right below the hairline. Now with the pads of your fingers of your left hand, search for the same area on the left side of your forehead. These pulsing points are the neurovascular holding points related to the pituitary gland.

By holding the pads of your fingers very gently upon these areas you can calm and soothe your emotional state of mind. Moments after contact is made, you may feel a slight pulsing in first one and then the other side from these points on your forehead. The pulsing eventually becomes synchronized on both sides.

It is essential to hold these points for 30 to 60 seconds in order for you to feel a calming and soothing effect. Although it is not necessary for you to feel the pulsing for this technique to be effective, most people become aware of the pulse after a few days of practice.

Your Personal Power to Control Your Addictions

After four weeks many of you will have developed the conscious ability to control your food addictions without using these techniques. At this stage you will be able to consciously shift from the craving sensation into the balanced sensation by willing it to be so. When you can do this, you will no longer need these anti-addiction activities. Your tendencies toward cravings and food addictions are finally under control. You have achieved your goal of having complete control over your food addictions. This activity is your first step in regaining this control over your addictions and your life.

The Short Version

Often you are in public places and will not want to draw attention to yourself by performing these activities. Since consistency is the key to success in this program, I developed a short version of the technique. It can be completed while sitting and attracts little, if any attention.

First - Activity #1

Quiet yourself for a moment and then picture in your mind's eye a scene where you are refusing to eat a particular food. Notice this instruction is different from the one in the long version of this technique.

After four weeks you will have the conscious ability to control your food addictions without using these techniques.

Anytime during your day, the anti-addiction techniques can also be used to replace the emotional effects of anxiety, stress and tension with the calming emotions of balance and peace.

Second - Activity #5
For a few seconds, lightly tap your adrenal pathway.
Third - Activity #6
For a few seconds, lightly tap your thymus pathway.
Fourth - Activity #7
Rest your finger pads on the right and left side of your temples for at least 60 seconds, while you continue picturing yourself refusing to eat the food which you are craving.

Releasing Anxiety and Tension

An anxiety is a fear or worry about some event that may or may not happen sometime in the future. Although the event is imaginary, it still has the ability to fire all of the Response Pathway Mechanisms of your body. This response to fear and worry has a devastating effect on a your ability to enjoy life. The Empowerment Patterning Technique described in my book, *Life Empowerment,* soothes and calms fears and worries as it subdues the emotional effects of anxiety and stress on the nervous system. The anti-addiction techniques described in *Eliminating the Stress of Weight Loss* are based on the techniques developed for the Life Empowerment Patterning Technique.

You can use the anti-addiction technique anytime you wish to replace the emotional effects of anxiety, stress and tension with the calming emotions of balance and peace. Follow the same directions but focus on your fear or worry while completing the activity.

112

CHAPTER EIGHT

ELIMINATING DIETING'S TOXIC WASTE BACK-UP PROBLEMS

THE STRESS OF TOXIC WASTE BACK-UP

The Problem of Toxic Waste Back-Up
Flushing The System
The 14-Day Cleansing Program
Using Acupressure to Eliminate Toxic Waste
Improving Skin Complexion While Cleansing Your Systems
Putting It All Together

This chapter teaches you how to eliminate the debris and waste products of fat cell destruction. We "lose" weight by destroying unwanted fat cells. The debris of this destruction can cause Toxic Waste Back-Up. This chapter gives you a 14-day cleansing program to help the elimination and purification systems of your body. It contains an acupressure treatment to aid digestion and elimination and, as an added bonus, it includes methods to improve skin complexion by removing even more of the toxic wastes from your system.

Notes:

Thoughts:

Actions:

Questions:

Additional References:

Tao of Balanced Diet by Dr. Stephen T. Chang
New Life Through Nutrition by Dr. Sheldon Deal
Acupressure's Potent Points by Michael R. Gach
Acupressure Weight Loss by Michael Reed Gach
Textbook of Medical Physiology by Dr. Arthur C. Guyton
Nutrition and Your Body by Benjamin and Sarah Colimore

CHAPTER EIGHT

Eliminating Dieting's Toxic Waste Back-Up Problems

The Stress of Toxic Waste Back-Up

The Problem of Toxic Waste Back-Up

One of the biggest saboteurs to a successful weight loss program rears its nasty head about two weeks after you start your diet. Successful weight loss means that the body is ridding itself of accumulated fatty deposits plus the debris and toxic substances that are stored in these areas. Depending on the condition of the dieter's digestive and elimination system, the excess amount of new factors that must be filtered, processed and excreted may increase to a level where the system cannot adequately handle the overage caused by the dieting.

The debris from dieting represents an increase in waste material that can clog the system. This blockage in your "drainage systems" forces all functions to back up and can cause major problems in a perfectly good diet where weight loss has been achieved. Within a few weeks your diet may lose its effectiveness, you may not feel as well as you did before starting the diet or worse, your body may start to retain and add weight, negating the prior weeks of successful weight loss.

Successful weight loss means that the body is ridding itself of accumulated fatty deposits plus the debris and toxic substances that are stored in these cells.

The importance of revitalizing the digestive functions during weight loss cannot be overlooked.

The importance of revitalizing the digestive functions during weight loss cannot be overlooked. You must allow for the proper flushing of the toxins and debris released when weight loss occurs.

The goal of this chapter is twofold. First it shows how to energize and revitalize those areas of the digestive system put under additional stress during weight loss. Second it shows how to drain and properly flush out the many waste products resulting from weight loss.

Flushing the System

Wars fought over water rights show that human beings always have realized that water is necessary for survival. If you are deprived of water, you will die in a short period of time. Yet, the daily consumption of pure drinking water remains one of the most neglected aspects of human nutrition.

The majority of Americans drink less than one glass of water a day. Yet, when I spoke to my patients, I noted that most of them believed that they were drinking plenty of water each day. Let's examine why this belief is prevalent.

It comes from the common misconception that other liquids like coffee, tea, fruit juices, milk etc., can replace water. You cannot substitute other liquids for water in your body any more than you can fill the battery of your car with milk or your steam iron with tomato juice.

People tend to believe that any liquid serves the same function as water. This is not the case.

When you work outside on a hot day, you lose a good deal of your body's moisture through perspiration and develop a tremendous thirst. Like most members of our society today, you quench your thirst with soft drinks, beer, iced tea, coffee, or imitation juices. However, pure water is needed to replenish the fluids in the system and to flush the body of toxic waste.

A healthy person needs to drink a minimum of six to eight glasses of pure water each day. During times of high stress or illness more water is recommended.

We are dependent on this fluid flow for good functioning of all the different systems of the body. Every living cell in your body requires water, just as it does nutrients and oxygen. Water acts as a transport medium in the body. The purer it is (free from minerals, softeners, and pollutants), the more nutrients it can carry to the cells. And, after depositing nutrients, this fluid is able to carry away waste products.

Never gulp your water. Take the time to slowly feel the water go down your throat.

Make a little ritual for yourself. Imagine that this is a wonderful, life-giving, renewing, cleansing liquid that you are swallowing. Then imagine this crystal clear liquid going through every cell in your body washing away all of the impurities. Next, visualize it flushing out residue and toxins, renewing your body tissues -- for it is doing all this!

You cannot substitute other liquids for water in your body any more than you can fill the battery of your car with milk or your steam iron with tomato juice.

When you drink adequate water, toxins released into the bloodstream from stored fat loosened by dieting are removed much more easily.

With each swallow, visualize your whole body responding and returning to its natural flow of vitality and life that you knew as a child. Bless the water on its way and bless your body for responding to the water so wonderfully.

Here's a simple but effective way to drink enough water each day. Save an empty quart jar that originally contained juice or some other fluid food item. After cleansing the jar thoroughly, fill it with good, pure room temperature water. Keep it with you if you spend your days away from home. Your goal is to drink all of the water before bedtime. This equals four to five glasses of water.

The 14-Day Cleansing Program

One of the major benefits of increasing the amount of pure water you bring into your body is the cleansing effect it has on your body. Toxins released into the bloodstream from stored fat loosened by dieting are removed much more easily and nutrients are able to travel more freely from one system to the next.

There are many programs you can use to enhance the cleansing and purifying processes. Before trying any program, check with your physician to make sure it is safe for you. This includes the two-part program that follows.

Part #1- The Master Cleanser
Cleansing Deep Tissues & Organs

The first part of this cleansing program was proposed by Dr. Stanley Burroughs. This method works well to loosen and remove stored toxins from deep tissues and organs, clearing out the toxins. It is simple enough for most people to do and it gives them an increase in their energy levels as well.

In a clean one-quart jar, combine the following ingredients:

> 8 tablespoons freshly-squeezed lemon juice
> 4 tablespoons pure maple syrup (no added sugar)
> 1/8 teaspoon cayenne pepper, and
> 32 ounces distilled water

You can adjust this ratio slightly to your taste. Sip frequently from this mixture throughout the day so that you finish it before you go to bed.

Part #2 - A Natural Laxative
Removing the Toxins from Your Body

It is important to keep the bulk in your intestines moving freely. To clear the loosened toxins generated through weight loss, take one teaspoon of cold-processed (unrefined and naturally processed) olive oil daily. Besides being a good nutrient, olive oil works as a liver tonic and intestinal lubricant.

Another option, which seems more appealing to many dieters, is drinking herbal teas. Many herbal teas are known

This cleanser works very well to loosen and remove any stored toxins from deep tissues and organs.

Digestion related acupressure points will, when stimulated, aid your digestion and assimilation.

for their laxative benefits. Some herbs you may look for are licorice root, anise, fennel or fenugreek. A daily cup of this type of tea aids in the toxic waste removal process. Your health food store is a good herbal tea source.

Using Acupressure to Eliminate the Toxic Waste of Dieting

The illustrations of Figures 8.1 and 8.2 mark the acupressure points you can use to eliminate the toxic waste of dieting. Refer to page 123 to see why we use acupressure. To activate these areas, gently rest your finger pads as close to the points as you can "guesstimate." Using the pads of your fingers, stimulate each point for a minimum of 60 seconds. Using approximately ten pounds of pressure, stimulate the area with a stationary circular movement, adding an occasional up and down motion. Feel free to stimulate each point up to two minutes.

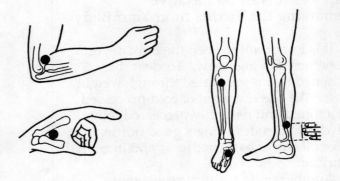

Figure 8.1
Acupressure Holding Points - Stimulate Both Right and Left Side of Body

Do not be too concerned about being on the "exact" spot. For up to two minutes apply this pressure over each broad two-inch area illustrated. You can approximate the position in the illustrations and feel assured that you have activated the correct area.

Figure 8.1 illustrates five digestion-related acupressure points found on both the right and left arm and leg. When stimulated, these points aid your digestion and assimilation. Search the general areas as seen on the diagrams. You may find that these areas are tender. This discomfort should disappear after a few sessions. Activate all points for a minimum of 60 seconds.

The five acupressure points illustrated in Figure 8.2 work to improve your elimination processes. Note that these points are all located only on your right side. Search the general area for the points shown in the diagram and stimulate each point with the same circular up and down motion. Continue stimulating each area for a minimum of 60 seconds.

These acupressure points work to improve your elimination processes.

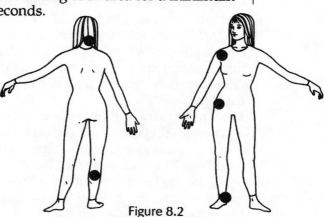

Figure 8.2
Acupressure Holding Points Located on Right Side Only

Each juice has a specific function which ultimately improves your skin tone and complexion.

Improving Your Skin Complexion While Cleansing Your Systems

Drinking fresh juices is a wonderful way to improve your skin tone and color but the ultimate purpose of drinking these juices is to cleanse your body systems.

While dieting, you can benefit from following this regimen three or four days each week for a 30-day period. Pick one of the following juices to drink each day.

I recommend drinking four glasses daily. Use the Biofeedback Reflex Test just to make sure that the juice you drink is not a stress-producing agent. Notice that each of these juices serves a specific function which improves your skin tone and complexion.

Carrot Juice - Carrot juice aids the liver and kidney function. It provides specific nutrients for the lungs and brain also.

Beet Juice - Beet juice has the ability to extract liver toxins, even those caused from years of abuse.

Green Juices - Green juices come from green-colored vegetables such as cabbage, spinach, celery, asparagus, etc., or any combination of green vegetables. These juices carry toxins across water membranes resulting in systematic flushing of the toxins.

Pineapple Juice - Pineapple juice contains a powerful digestive enzyme called bromelain which cleanses the small intestine.

Comfrey Tea - Comfrey tea has been used for centuries to heal and soothe the stomach lining. It also stimulates the production of the enzyme pepsin.

Putting It All Together -
The Overall Cleansing Program

Follow this four step cleansing program for a minimum of 14 days. For maximum benefit it should be followed for 30 days.

Complete each procedure every day.

Step 1
Prepare one quart of the master cleanser and sip it throughout the day (page 119).

Step 2
Three or four days each week, drink fresh juice throughout the day (page 122).

Step 3
Drink one cup of laxative tea **OR** take one teaspoon of naturally processed, unrefined olive oil each day of your diet (page 119).

Step 4
One time each day, massage the accupressure points that aid digestion, assimilation and appearance (pages 120 and 121).

Why Use Acupressure?

Some programs in this book incorporate the use of acupressure to accomplish change. The art and science of acupressure originated from the meeting of three separate paths -- curiosity, observation and trial and error.

Since the beginning of recorded history, human beings have tended to hold places

This cleansing program eliminates the dieter's problems of toxic waste back-up.

Wherever you are, you can facilitate healing processes by using mankind's original medical tool - your hands.

on their body that ache or hurt. To help relieve excessive pain, we are drawn to touch or hold injured areas of our body. This is a basic form of acupressure. When you place your hand on your forehead to help relieve a headache or hold your bruised leg to lessen pain, you are actually treating yourself with basic acupressure holding points.

Throughout the ages, people have found the most effective ways to help themselves through trial and error. Hundreds of years of this objective observation led to the conclusion that certain problems could be relieved by stimulation of specific nerve points located on the surface of the body. Effective methods were shared through the centuries within families and in schools and were integrated into the system we now call acupuncture. The treatment methods were first put on permanent record thousands of years ago in the book, *The Yellow Emperor's Classic Internal Medicine.*

There is a dual benefit derived from doing acupressure on yourself. The first and most obvious is, by assuming responsibility for your successful treatment, your self-confidence and trust in yourself greatly increases.

The second benefit is that, once learned, this self-treatment knowledge is with you for the rest of your life. Wherever you are, you can facilitate healing processes by using mankind's original medical tool — your hands.

CHAPTER NINE

ELIMINATING OLD NEGATIVE BEHAVIORAL REACTION PATTERNS

RELEASING CONDITIONED RESPONSE MECHANISMS

Addicted to Negative Behavior
Programmed to Fail
Changing Old Behavioral Patterns
The Solution
The Process
Releasing Your Hidden Splendor
Releasing Conditioned Stress Responses
Changing The Inner Voice
Daytime Conversations With Your Subconscious Control
The Healer Within

This chapter teaches you how to release negative behavioral patterns that were built into your subconscious memory. Stored deep in your subconscious memory are many old, negative behavioral reaction patterns that you no longer need in your present day life. These memory patterns create poor self-confidence and negative behavior problems that seem to surface for no apparent reason. To successfully achieve and maintain weight loss goals, ending this cycle is mandatory.

Notes:

Thoughts:

Actions:

Questions:

Additional References:

Creative Visualization by Shatki Gawain
You Can Heal Your Life by Louise L. Hay
Mind as Healer, Mind as Slayer by Kenneth R. Pelletier
The Psychology of Consciousness by Robert E. Ornstein
The Physiopathology and Treatment of Functional Disorders
by Dr. George Whatmore and David Kholi

CHAPTER NINE

Eliminating Old Negative Behavioral Reaction Patterns

Releasing Conditioned Response Pathway Mechanisms

"Mind is the master power that molds and makes, and man is mind, and evermore he takes the tool of thought, and, shaping what he wills, brings forth a thousand joys, a thousand ills: he thinks in secret, and it comes to pass: environment is but his looking-glass."

James Allen
As a Man Thinketh

Addicted to Negative Behavior

One of the most powerful saboteurs that can ruin the success of your weight loss program is the result of old negative behavioral reaction patterns stored in your sub-conscious memories. How you have been conditioned to think about how your body looks, shapes your self-image and self-respect. How you think about your body has a direct bearing on your actions with respect to it. Your subconscious beliefs and your feelings about your body influence your eating and exercise habits for good or for ill.

Without being consciously aware of it, you may be sabotaging your weight loss goals by maintaining earlier negative reactions that have become a conditioned

Without being consciously aware of it, you may be sabotaging your weight loss goals by maintaining earlier negative self-programming.

127

The first step in any weight loss program must consist of making the conscious choice to direct your mind to cooperate with your weight loss goals.

response. You may subconsciously believe that you do not deserve to be thin. You may still retain the "fat little kid" image you developed when you were younger. If these negative thoughts and images exist, you will be unable to change other more influential patterns. This is why it is so difficult to change or maintain your weight loss goals through diet alone.

The first step in any weight loss program must consist of making the conscious choice to direct your mind to cooperate with your weight loss goals. Then you have to learn how to channel your mind power. To do this successfully, it is necessary for you to understand the process your mind uses to manipulate your body.

Your self-image is built up of a combination of emotional memories that are stored in your autonomic nervous system. These memories are the sum total of your personal experiences and all of the remembered feelings you associate with them. These old, stored voices and visual images in your brain can be triggered by present day feelings. This can be either positive or negative. It is negative when they cause you to re-experience those old feelings and emotions that have created your present weight problems. These re-activated thoughts in your mind, may be the taunts of little children calling you fat or the voice of your parents referring to you as chubby. This makes you "feel" fat and chubby now.

When this happens these conditioned negative behavior patterns, your old childish responses,"kick in" and take over.

Automatically, with no conscious thought, you find yourself eating something which, for some unknown reason, "makes me feel better."

In Chapter One, I discussed the phenomena whereby such factors as fatigue, stress, illness or Unsuspected Stress Foods trigger long-stored, conditioned behavior responses. External triggers that reactivate old conditioned behavior patterns cause distress. Until someone or something helps you eliminate incorrect triggers, without conscious thought, you revert back to these old, stored conditioned behavioral responses, even though they are being triggered by the wrong stimulant. It is this mechanism that will defeat even the best of diet plans.

Diet programs fail when the dieter reverts back to a conditioned negative behavioral pattern developed in response to earlier life experiences. Their diet is sidelined by any of their present-day stressors that mistakenly activate these old behavioral patterns.

Programmed to Fail

The basic nature of human relations requires all of us to be raised by parents who teach us acceptable behavioral patterns. In our society almost all social action is learned behavior. As a child grows up, he is continually corrected and continually told "no." Under the strict guidelines of a family upbringing, by the time this child is eight years old, he may have the feeling that he "cannot do anything right". This

Diet programs fail when the dieter reverts back to a conditioned negative behavioral pattern developed in response to earlier life experiences.

Psychologists have shown that 90% of us have a problem with our self-image by the time we are ten years old.

feeling is compounded for a child who has a weight problem. He may be dealing with endless taunts about his weight from a sibling or from other children in the neighborhood. Add to this the relatives who lovingly tease the child because he is chubby and you have a child who has a poor self-image.

It's a common situation. Psychologists have shown that 90 percent of us have a problem with our self-image by the time we are ten years old. This poor self-image problem leads to poor self-confidence, which in turn leads to setting low objectives. When our theoretical child sets low objectives, he sets himself up to get poor results and low productivity. This reinforces the initial problem of poor self-image setting in motion a vicious cycle (Figure 9.1). At this point the child is programmed to fail. A conditioned neurological pathway is now set up in the brain and nervous system which guarantees this cycle's existence.

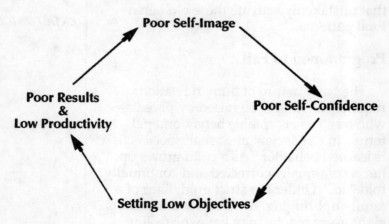

Figure 9.1
The Self-Fulfilling Prophecy of Poor Self-Image

Changing Old Unnecessary Conditioned Programmed Behavior

Negative conditioning and perceptions of the mind must be corrected in order to insure both a successful weight loss program and then maintaining hard-earned weight loss goals. Learned negative behaviors must be abandoned in favor of new behavior patterns. This calls for changes to be made.

Psychotherapy (treatment limited to interview and discussion techniques) appears to have little correcting affect on this type of behavior pattern. Research demonstrates that patients who have undergone prolonged psychoanalysis and other forms of psychotherapy show no significant reduction in their ability to control old conditioned reaction patterns. In fact, a person treated with psychotherapy alone can develop more problems. The procedure uncovers materials to which the individual might possibly be reacting, but it does not directly alter the conditioned reactions to this material. Some forms of psychotherapy reduce the stress value of a troublesome factor in a person's life without altering their response, hence leaving them vulnerable to future distress.

Many psychotherapeutic failures are due to the presence of an unrecognized, hence un- treated, improper response pathway (the involuntary use of the wrong conditioned reaction to a non-specific event such as the slap in Chapter Two). It is not necessary to dredge up or invest in lengthy psychotherapy to replace this patterned behavior. What must be done is this:

Negative conditioning and perceptions of the mind must be corrected in order to insure a successful weight loss program.

You must rewrite the dialogue that exists in your mind and re-draw a new mental picture.

Step 1
You must change the voices and visual images that are stored in your mind.
Step 2
You must rewrite the dialogue that exists in your mind and re-draw a new mental picture.
Step 3
You must have a way to tie into the subconscious mechanisms that are at play here.
Step 4
You need a method to introduce a new "script" to that subconscious mind which is responsible for creating a new self-image.

The Solution

Modern technology has given us the tools to help release the negative self-images that cause so much individual failure. With these tools, you can reclaim genetic potentialities by reprogramming the subconscious mind.

Researchers in the field of biofeedback have learned how to contact and release the power of the subconscious. They discovered that the subconscious mind is opened for suggestions right before entering into the drowsy, sleeping state. It is during this "window of opportunity" that you can reprogram old patterns and replace any negative programming that is causing problems.

You accomplish this reprogramming by feeding your subconscious mind new concepts and ideas to replace the old ones. For the next 30 days, each night before you fall asleep, you will listen to a recording that you have personally made. This tape recording contains your creative description of the self you desire to become. These new visual pictures will replace negative programming stored in your subconscious mind. This technique helps to release built-up anxieties and tensions. It can be used to reinforce an image of the slimmer, more attractive you. It will release those unnecessary conditioned response reactions that are causing so much difficulty in your life.

Over a 30-day period your subconscious mind will accept these new pictures and eventually replace old negative patterns. You will be replacing the negative patterns with enriching, empowering pictures and perceptions. You will find that a great many of your problems will seem to disappear be- cause of this directed conscious communication with your subconscious control mechanisms. You will gradually replace the negative self-images that you have carried around all your life with new, positive self-images.

Although there are taped messages available commercially, my research shows that many people respond poorly to an unknown authority figure telling them what to do. Just as you rebelled when your parents told you what you could or could not do, you resist the messages of these tapes. For you to get the true benefit of this technique, you must listen to your own voice telling you how you would truly like to live your life. Your subconscious mind listens to you telling yourself what to do much better than it listens to someone else.

It is during this "window of opportunity" that you can reprogram old patterns and replace any negative programming that is causing problems.

You will be replacing the negative patterns with enriching, empowering pictures and perceptions.

When you plant a seed in the ground, you do not dig it up all the time to see what is happening. You let it grow and take root and eventually you step back and look at the results of this growth process. Follow the same approach with this tape process. Use this technique for 30 days to allow your subconscious to receive the benefit of this effort. After 30 days, feel free to make a new tape or continue with the one you have. After a few months you may have a variety of tapes. Depending on how you feel on a specific day, you will be able to choose the best message for that day.

The Process

To create your new self-image you have to construct a story from the four scripts that follow. Only choose phrases that sound and feel true to your inner self when played back to you. There are four categories of scripts. When creating your new self-image tape, you must read from all four scripts. You may include as much of each script as you like but do not read any phrase that does not "feel good" to you.

The time you spend presenting this new self-image tape to your subconscious mind will become a very special time for you. During this time you are free to stimulate your mind to see your life and thinking patterns more clearly.

This is your special time to fill yourself with the truth of the splendor of your uniqueness and to see yourself as a beautiful light, a diamond of perfection and beauty.

Releasing Your Hidden Splendor

To make your customized bedtime story, record yourself reading from all four scripts that follow. Your tape should be no longer than 25 to 30 minutes.

Step 1
Read through all of the following scripts. You will record the entire text for Script #1 and Script #4. For Scripts #2 and #3, use a pen or pencil to cross out those phrases that you would not "feel comfortable" saying or hearing.

Step 2
Plan a quiet time when you can sit down and relax with your tape recorder and all four scripts in front of you. Both the energy and enthusiasm expressed in your voice are important. Make your tape when you are feeling your most calm, positive and happy.

Step 3
The tone of your voice while you read your scripts should be smooth, calm, relaxed, yet firm. Do not rush. Pause for at least three seconds after each sentence or thought so that your thought mechanisms have a chance to actually picture and savor the concept you are presenting.

Step 4
If possible, use a tape machine with an automatic shut-off, for it is likely, and OK, for you to fall asleep while listening to the tape.

This is your special time to fill yourself with the truth of the splendor of your uniqueness and to see yourself as a beautiful light, a diamond of perfection and beauty.

*I am free.
I will
awaken
with the
wonderful
feeling and
inner peace
of knowing
that all this
and more
is waiting
for me.*

Step 5
Turn on your tape recorder. Introduce yourself with the following statement: "Hello, this is (your first name). It is time now to close my eyes. I am going to take three deep breaths and exhale the air slowly through my mouth." You can talk to yourself using the noun "I" and there will be no confusion between the I that is listening and the I that is speaking.

Step 6
Slowly and in a calm relaxed voice say, "I am going to count backwards from ten to zero. I can see the numbers appear on a blackboard as I hear them." Proceed to count backwards from ten to zero.

Step 7
Slowly and in a calm voice read out loud from all four categories. Read them in their order of appearance.

Step 8
When you finish your taping, read out loud the following statement: "And now I am free. I will awaken with the wonderful feeling and inner peace of knowing that all this and more is waiting for me to experience today. I give thanks for my great blessings and accept my total being in its perfection and unity."

Releasing Conditioned Stress Responses

It is very difficult, if not impossible, for the average person to consciously release the muscle response pathway that causes both muscle bracing and muscle tension. Listening to your tape with the intent of hearing the phrases from Script #1 facilitates this desired response. It does this by relaxing your muscles and releasing tension. Also, your circulation and respiration response pathways will begin to return to a non-addictive state. Listening to the first script calms and relaxes you and you will be better prepared to receive the full benefit of the next three scripts.

When making your recording, be sure you pause for at least three seconds after each phrase or statement so that you can picture what you are hearing in your "mind's eye". Please use this entire script.

Script #1
"It is now my time to relax. It is my time to rest and relax my eyes. I am now taking a long deep relaxing breath. I am ready to listen, feel and experience these beautifully relaxing images."

(Repeat each phrase two or three times)

"I feel my toes warm and relaxed. / I can now feel pulsing warmth in my toes. / I feel my ankles warm and relaxed. / I can now feel pulsing warmth in my ankles. / I feel my right leg warm and relaxed. / I can

Picture what you are hearing in your mind's eye.

Imagine that you are telling a wonderful story with a happy ending to a small child.

now feel pulsing warmth in my right leg. / I feel my left leg warm and relaxed. / I can now feel pulsing warmth in my left leg. / I feel my low back warm and relaxed. / I can now feel pulsing warmth in my low back. / I feel my mid-back warm and relaxed. / I can now feel pulsing warmth in my mid-back. / I feel my shoulders warm and relaxed. / I can now feel pulsing warmth in my shoulders. / I feel my neck warm and relaxed. / I can now feel pulsing warmth in my neck. / I feel my arms warm and relaxed. / I can now feel pulsing warmth in my arms. / I feel my hands warm and relaxed. / I can now feel pulsing warmth in my hands. / I feel my face warm and relaxed. / I can now feel pulsing warmth in my face. / My entire body is now relaxed. / My mind is now quiet and still. / My mind is now quiet and happy. / I am at peace."

Changing the Inner Voices and Visual Images Stored in Your Mind

Thoughts are energy that can be built up into complex images and actually seen in the mind and felt in the muscles. In this section you will provide your subconscious mind with the necessary images to establish a more positive self-image.

Take a few minutes to read all of Script #2. You will notice that each phrase is filled with wonderfully creative images. Cross out any phrase that you do not like or that does not "feel good" to you.

You may enjoy having all of these phrases in your first recording and in later recordings, become more selective.

When making your tape, read slowly and feel free to repeat any phrase as many times as you wish. Imagine that you are telling a wonderful story with a happy ending to a small child. Create a beautiful fantasy. Fill yourself with color, light and energy. Give yourself warmth if you are cold and cool breezes if you are warm. Have water soothe your aches and pains and let sunlight fill your heart. Crown your head with stars and let moonbeams shine on your weariness and inspire you to dream. Let the sea wash away your fears. Use jewels to restore your energy. Have trees give you strength and let beautiful fields of wild flowers enrich your life experiences. Give yourself magical gardens to heal you with their perfumes. Allow waterfalls to wash away all negative thoughts and rainbows to fill you with joy and divine love.

When making your tape recording, make sure you pause for at least three seconds after each phrase or statement so that you can create and enjoy each beautiful picture in your "mind's eye".

Fill yourself with sunlight to fill your heart, with moonbeams to inspire you to dream and with rainbows to give you joy.

Script #2

"I am surrounded by a robe of glory. / I feel beautiful. / I see my beauty reflected in every flower. / I am a waterfall. / I am one with the harmony of nature. / I am as the perfect simplicity and beauty of a daisy. / I attract beauty and love as the nectar attracts birds and bees. / My brilliance is revealed. / My life is harmony. / I am embracing my world with my loving. / My life is becoming one of total harmony and peace. / I feel

*My way
is now
total
harmony
and
peace.
I feel love.
I am love.
Love is all
around
me.*

love in my heart. / I am love. / Love is all around me. / My life is surrounding me with peace, love and harmony. / I am blessed with the true gifts of nature's harvest. / I am free to choose everything for my highest good. / I can now accept the gifts of the bounty of the earth. / I am aware of the endless supply of beauty that is flowing into me now. / Old feelings of confusion are gone now. / I am achieving health and vitality. / Whatever I need will be supplied. / The power of purification flows through me. / Fantastic strength and vitality is coursing through my body. / I am moving through the waters of life. / I am receiving total rejuvenation. / I am cleansed and purified. / I am the fountain of youth. / I am revitalized. / I am experiencing my hidden splendor. / I am being reborn. / My self is changing. / I am awakening from the past. / I see myself as a shining star. / My starlight is all encompassing. / My starlight is everywhere. / I can now master the worlds of my being. / There is true eternity in this moment. / Life is eternal. / My will is strong. / I am being guided through this process of reconstruction. / My energy is transformed. / I am creating a new life, a new way, a new me. / The unnecessary drops away. / New energy fills me now. / The rebuilding will be successful. / My prayer is answered. / I am breaking free from old patterns. / Darkness is draining away from me now. / I am released. / Swirling, healing energy is flowing through me. / It feels warm and gentle. / I feel myself growing strong. / There is a pure white light that radiates around me protecting me. / I am whole. / I am free. / I am beautiful."

Expressing Whatever You Would Truly Love to Achieve

This is your opportunity to create new choices for your subconscious control mechanisms.

To affirm means to make firm. An affirmation is a strong positive statement that something is already so. It is a method of presenting new, positive visions of the "you" that will make your life work even better. Allow the wonderful thoughts that you are about to read to replace old negative thoughts and images that are holding you back. This is your opportunity to present a new vision of your life to your subconscious control mechanisms.

Recognize the importance and strength of each of these affirmations. This is a truly powerful part of your taped message. If you find one affirmation that you want to repeat many times, please feel free to do so.

Choose affirmations that will help you release your problems and give you the strength to face tomorrow's situations. Select affirmations that make you feel good and know that you deserve to feel this good. Reading the entire section is beneficial. Do so if you can.

Read slowly and pause for at least three seconds after each phrase to allow your "mind's eye" the opportunity to savor the importance of each phrase. Repeat the affirmation as many times as you like.

Remember to read each affirmation as a positive statement, as something that is already a fact.

I have radiant love within me.

Script #3

"I enjoy my own company. / I am free to succeed in every effort. / I am at ease with myself and with others. / I am free of guilt and blame. / I am capable and confident. / I am free to change. / I am free to be cheerful. / I can easily forgive others. / I forgive myself. / I am free to love and forgive myself. / I have radiant love within me. / Divine love is healing me now. / My mind is quiet and still. / I am free to feel loving compassion for everyone. / I am filled with loving kindness. / My heartbeat is calm and regular. / My body breathes freely and comfortably. / All parts of my body are being filled with healing energy with every breath I take. / I am free to enjoy sufficient energy to accomplish my desires. / I am free to enjoy sufficient wisdom to accomplish my desires. / It is OK for me to have fun and enjoy myself. / I am free to smile. / I smile freely. / I am free to have fun. / I am relaxed and centered. / It is OK for me to enjoy everything that I do. / I give thanks for divine restoration of my mind. / I give thanks for divine restoration of my body. / I give thanks for divine restoration of my financial affairs. / I give thanks for divine restoration of all my relationships."

A Happy Ending

Every good story must have a happy ending. Script #4 ties all of the thoughts of the first three scripts together. Remember to read the entire script slowly to give your "mind's eye" a chance to visualize and enjoy each phrase.

Script #4
"The spirit of life which is active in me, flows through my physical body in a purifying, cleansing, healing stream that removes all obstructions and brings peace, health, beauty and harmony to my body. / I am well, strong and vital. / I am peaceful and poised. / I am buoyant, happy, free and beautiful. / I shall awaken filled with energy, radiance and the power to accomplish whatever I choose to do. / I give thanks that my life forces are now conserved and controlled. / I give thanks for the orderly adjustment of every function of my body. / I am strengthened, healed and renewed. / Divine love is healing me now."

To finish releasing your hidden splendor, end your tape with the following statement from Step 8 (Page 136).

"And now I am free. I will awaken with the wonderful feeling and inner peace of knowing that all this and more is waiting for me to experience today. I give thanks for my great blessings and accept my total being in its perfection and unity."

I shall awaken filled with energy, radiance and power to accomplish whatever I choose to do.

At any time on any given day, stress, anxiety, tension or worry may trigger those old stored visual images and pictures that surface to sabotage weight loss goals.

Daytime Conversations with Your Subconscious Control Mechanisms

At any time on any given day, stress, anxiety, tension or worry may trigger those old stored visual images and pictures that surface to sabotage weight loss goals. Use this opportunity to erase these old stored voices forever. Replace them with new, empowering, strengthening voices and images that will help you achieve your goals. The following script provides a daytime message to these subconscious mechanisms. This soothing, healing and restoring voice will shut out and erase your negative response reactions.

Take a few minutes to read all of the following phrases. Cross out those you do not want in your tape. Highlight those that you may want to repeat a few times. Then sit down with your tape recorder and in a strong, positive and relaxed voice, make your daytime tape. Remember, those old voices stored in your subconscious memories were created by other people telling you what they wanted you to believe - and you bought it. This is your special time to create new, more powerful voices and images that will create a happier, healthier and more successful life for you. Listen to this tape to change your life anytime during the day when you are feeling discouraged or depressed (personal earphones are recommended). Remember, you deserve to enjoy a happy, healthy and successful life.

"I am free to change. / I am willing to change. / I digest new ideas easily and effortlessly. / I love and approve of myself. / My future is bright and secure. / I am moving forward in my life with great success. / I love life. / I am free to love myself. / Success is my birthright. / My life moves forward toward my highest good. / I am free of negative thoughts. / I am enjoying my life in perfect ways. / My body is in a constant state of healing. / My past is forgiven. / Miracles are created in my life daily. / It is my divine right to create my own direction in life. / I am safe. / I am free. / My mind knows my true identity. / My heart knows my true purpose. / My body knows my true function. / I am moving lovingly and confidently forward. / I am at peace. / I move from the old to the new with ease and joy. / My needs will be met. / I am trusting in the process of life. / I am tolerant of myself and of others. / I forgive and release all past experiences. / I release all criticism of myself. / I release all criticism of others. / A wonderful new future lies before me. / I am free to express myself openly. / My creative expression is free. / I am guided to my success- filled future. / It is safe for me to ask for what I want. / It is safe for me to get what I want. / It is safe for me to be who I am. / Only good comes from my wants and desires. / I can easily ask for what I want. / All that I need is generously and abundantly supplied to me. /

I am moving lovingly and confidently forward.

145

I am replenishing myself with nurturing experiences.

I am a deserving person. / Life supports me. / I am divinely guided. / I am deserving of love and happiness. / Joy fills my heart. / My vulnerability is my strength. / My openness is my guiding light. / My truth leads me to the best choices. / I am surrounded by love. / I am supportive of my loving nature. / I am at peace in every area of my life. / I am capable. / I am free of doubt. / My tasks are easy to accomplish. / I am breathing freely, fully and completely. / I release all unhappiness. / There is plenty of time for me. / There is plenty of time for all I wish to accomplish. / I choose to make my life light and easy. / Old patterns are dissolved. / I am replenishing myself with nurturing experiences. / My future is open to natural creative expression. / I attract loving, prosperous, and supportive friends. / It is natural for me to enjoy myself. / Success comes to me in perfect ways. / I choose to surround myself with beauty. / I am flexible and I can change easily. / I am confident of my success. / I am safe in trusting the process of my life. / I am safe. / I love myself. / I love others. / I lovingly approve of myself. / I am confidently caring for my well-being. / I am creating harmony in my life. / I am wanted. / I am welcome. / I am deeply loved. / Each moment of my life is perfect. / I am relaxed. / I am breathing comfortably. / My life flows comfortably toward my goals. / Creative ideas manifest divinely. / My life is guided sweetly and joyously forward. / I am the authority in my life. / I am the author of my life. / My decisions manifest successful results. / My decisions manifest loving results. / My opinions are valued by me and by others. / I speak gently and lovingly. / I listen patiently and clearly. / I approve of myself. / I choose to handle all of my daily experiences with patience and ease. / I attract loving and supportive experiences. / I communicate clearly with love. / I trust my intuitive abilities. / I am my own security. / Joy now flows freely within me. / I am at peace with life. / I am wonderful. / My mind and body are in perfect balance. / I am a powerful and dynamic person. / Every part of my body is functioning perfectly. / My spirit is forever young. / My decisions are right for me. / I approve of myself and others. / My experiences get better and better. / I am free to express the joy of living. / I allow myself the freedom to totally enjoy every moment of every day. / I am the living, loving, joyous expression of life. / All is well in my world."

Recognize that a very special healing mechanism exists within your autonomic nervous system.

Listening to these new voices and visual images for 30 days will create a new more positive you.

The Healer Within

Recognize that a very special healing mechanism exists within your autonomic nervous system. It is this internal healing mechanism that gives you the ability to rejuvenate, revitalize and regenerate your body, mind and spirit. This mechanism exists for your benefit. The degree to which you allow it to help you depends on your own volition to maintain a high degree of awareness about your body's reactions to your weight loss goals.

Listening to these new voices and visual images for 30 days will create a new more positive you. It is the best tool to help prevent any old unnecessary behavior reactions from sabotaging your weight loss goals or any other of your life goals.

ELIMINATING FACIAL LINES & WRINKLES CAUSED by DIET & STRESS

THE COSMETIC ACU-POINT FACIAL TONING PROGRAM

The Stress of Weight Loss and
Its Effects on Facial Lines and Wrinkles
Facial Stress and Tension
Erasing Lines and Wrinkles
Finding the Cosmetic Facial Acu-Points
The Cosmetic Acu-Point Facial Toning Program
Your Healthy Skin

Chapter Ten presents The Cosmetic Acu-Point Facial Toning Program of The Monterey Wellness Center. Using this program gives an added boost to your weight loss plan because it enhances facial beauty. It teaches you how to use acupressure to tonify the underlying muscles of the face. Many books teach this toning program as a facelift program.

Notes:

Thoughts:

Actions:

Questions:

Additional References:

The Acupressure Facelift by Dr. Kurt Donsback
The Acu-Facelift Booklet by Michael Reed Gach
New Beauty - The Acupressure Facelift by Lindsay Wagner
*Acupuncture: The Ancient Art of Healing and How it Works
Scientifically* by Dr. Felix Mann

CHAPTER TEN

Eliminating Facial Lines & Wrinkles Caused by Stress & Diet

The Cosmetic Acu-Point Facial Toning Program

The Stress of Weight Loss and Its Effects on Facial Lines and Wrinkles

Most people who diet are primarily interested in improving their appearance. But, because weight loss often tends to accentuate facial lines, this book includes a method of facial toning that provides two things: the erasure of lines and wrinkles of facial skin and the long range benefit of tonifying the under-layers of skin to help prevent new wrinkles from forming.

During rapid weight loss, the removal of fat cells that lie near the surface of the skin can leave you with more lines and facial wrinkles than you had before starting the program. Although this condition is temporary (a few days to a few weeks), the facial toning program in this section teaches you a way to correct this situation immediately. Unlike astringents and other solutions that offer temporary correction for a few hours, these techniques provide fast and permanent toning of facial skin.

Rapid weight loss can leave you with more facial lines and wrinkles than you had before starting your weight loss program.

Since they first appeared in print 5,000 years ago in *The Yellow Emperor's Classic Internal Medicine,* the techniques taught in this chapter have been used to eliminate facial lines and wrinkles.

In this chapter you will learn to locate and stimulate the 24 acupressure points located on the face and neck. Self-treatment of these points eliminates lines and wrinkles that have already formed around the mouth, eyes and other areas of the face. More importantly, this treatment prevents the formation of additional premature facial lines by rejuvenating fatigued skin and underlying tissue.

You can perform this self-given treatment anytime during the day as long as you are not severely fatigued. You may prefer to give yourself the treatment in the evening prior to bedtime.

Figure 10.1
Smooth Skin Surface

Erasing Lines and Wrinkles

To illustrate how lines and wrinkles are created on your face, refer to Figures 10.1, 10.2, 10.3 and 10.4. They show a crosscut of the surface and underlying tissue of facial skin.

Figure 10.1 demonstrates the youthful skin with all layers (upper, middle and deep muscle) at their proper tone and tautness.

Wrinkles and age-revealing lines appear when the layers start losing their suppleness and elasticity. Facial lines and wrinkles are the result of crevices that develop all the way down to the lowest layer of skin (10.2).

Figure 10.2
Lines & Wrinkles Form

Cosmetic creams temporarily tighten the upper layer of skin, leaving underlying tissue layers untouched (Figure 10.3).

The deep effect of cosmetic acupressure completes the rejuvenation of facial skin. It permanently erases lines and wrinkles down to the deepest levels by making all layers of the skin taut (Figure 10.4). It reverses the aging process and results in a more youthful looking face.

Why Use Acupressure?

Almost all physical problems of the human body manifest themselves through increased muscular tension or stress. This muscular tension inhibits the body's ability to effectively cope with a disrupting condition. Acupressure enhances the body's homeostatic process by directly reducing muscular tension and stress.

Figure 10.3
Surface Toning Only

This self-treatment smoothes facial stress lines and stress wrinkles because muscular stress tends to accumulate around the acupressure points. By treating these points on the face and tonifying the skin, you will not only remove lines and wrinkles by tonifying the skin, you will remove lines and wrinkles caused by stress and tension as well as the ones created from rapid weight loss.

These acu-points are easy to locate and stimulate. Using your finger pads, feel for a tender spot in the approximate location of the acu-point (Figure 10.7). Hold your finger pads directly on this area and make stationary circles for 40 seconds as you gently press with your finger pads.

Figure 10.4
Full Acu-Toning

Acupressure is a method of using pressure on nerve endings that are close to the skin surface.

The Application of Pressure

Acupressure is a method of using pressure on nerve endings that are close to the skin surface. Although a medium pressure is recommended for this procedure, the effective amount of pressure to use varies for each person. You should feel the pressure of the stimulation, but you should not feel pain. Most people press too hard for the first few days. If you do this you may find that the points have become tender. If this happens, use less pressure or discontinue the treatment until the tenderness subsides.

Press the point in a small (pea sized) stationary circular movement with an occasional up and down motion (Figure 10.5). Treat each point bilaterally, starting with the top right and left acu-points and continuing this bilateral treatment method until you have worked your way down to the bottom points. Apply pressure on each set of points for 30 to 40 seconds. It is not necessary to treat the points for longer periods. Increased stimulation does not produce greater results.

Treatment Time

Until you are familiar with all of the points, you may need six to ten minutes to stimulate all of the cosmetic acu-points for facial toning. As you become more familiar with the points and how to stimulate them, you will find you need less time.

The Treatment Tool

Some books suggest using the index finger to stimulate and treat each point. I recommend that you start off using the pad of your pinky or fourth finger for the first few days. These two fingers are generally not as strong and you will not be tempted to use too much pressure.

To prepare your skin for this acupressure treatment, you need to stimulate the nerves and blood supply.

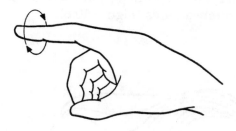

Figure 10.5
The Treatment Tool

The Preliminary Stimulation

To prepare your skin for this acupressure treatment, you need to stimulate the nerves and blood supply. You do this by gently patting the face and throat. This helps stimulate the underlying muscles so they can accept the full benefit of your acupressure treatment.

This helps stimulate the underlying muscles so they can accept the full benefit of your acupressure treatment.

Start from the base of your neck. Using the pads of your fingers, gently but vigorously, pat your skin from the base of your neck to your jaw bone. Do both sides at least three times. Next, using the pads of your fingers, gently but vigorously pat your skin, starting from your jaw bone and proceeding up your cheeks to your eye level.

Do not pat your eyes! Do each pathway at least three times. Then tap your forehead area. Use the finger pads of both hands to make a typing style of tapping across your entire forehead area (Figure 10.6).

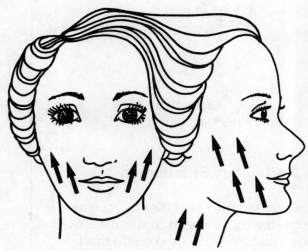

Figure 10.6
Facial Stimulation

How to Find the Cosmetic Acu-Point Facial Toning Points

Figure 10.7 illustrates the locations of the cosmetic facial acu-points. The more accurately you locate a point, the stronger the effects of its stimulation. As you probe around, you will find that the right spot is more sensitive to pressure than the surrounding area.

The more accurately you locate a point, the stronger the effects of its stimulation.

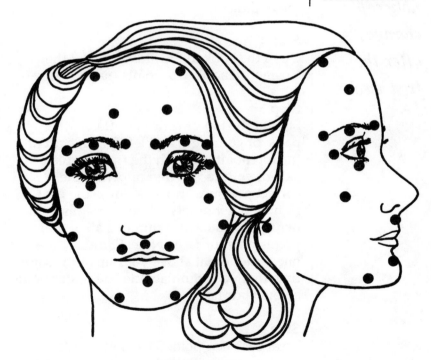

Figure 10.7
The Cosmetic Facial Acu-Points

The Cosmetic Acu-Point Facial Toning Program will begin to make noticeable changes after the first week.

The Cosmetic Acu-Point Facial Toning Program

Step 1
Wash and clean your face.

Step 2
Apply your normal skin conditioners.

Step 3
Use the patting technique to gently stimulate the face and throat for two minutes. This prepares the underlying muscles for the treatment.

Step 4
Open the book to the illustration of the Cosmetic Facial Acu-Points (page 157).

Step 5
Start at the top of your face. Using the pads of your right and left finger, start at the top points on both the right and left side of your face. After applying the correct pressure, move down to the next right and left points. Proceed in this manner until you have stimulated all of the points on your face. Note: Stimulate the one center point above your lip at the same time that you stimulate the one center point below your lip.

Step 6
Treat each acu-point 30 to 40 seconds. Gently press with the finger pads in circular massaging motions with occasional up and down fluctuations.

Document Your Results

It is human nature to forget how bad things were after they begin to improve. This causes discouragement in both weight loss and facial improvement programs. Changes that last are always slow to develop. The Cosmetic Acu-Point Facial Toning Program will begin to make noticeable changes after the first week. Before you start the program, I recommend that you take a close-up picture of your face (with no make-up). Retake the picture after 30 days, wearing the same clothes in the same setting. This comparison picture will truly lift your spirits and encourage you to continue with this program. After your initial 30-day treatment, a two-time-per-week maintenance program prevents new lines and wrinkles from forming.
Caution: Please use common sense. Never work on your face if you have serious skin problems, infections, eczema or rashes. Consult with your physician if you have any questions.

Your Healthy Skin

The manner in which your internal organs are functioning greatly influences the tone, texture and color of your facial skin. Throughout the ages, health spas have recognized this and have used self-care programs similar to the ones outlined in Chapter Eight. You can enhance the effects of this facial toning program if you follow the 14-day revitalization program outlined on page 123.

This comparison picture will truly lift your spirits and encourage you to continue with treatment.

These self-care programs enhance the Cosmetic Acu-Point Facial Toning Program.

While completing this facial program, allow your skin to be naturally moist. Over-exposure to sun and wind causes the skin to become leathery and dry. You can correct this condition by applying lemon peel, vitamin E or freshly-cut cucumber slices to the face to safely and effectively soothe, heal and rejuvenate your skin.

During the years that I have been teaching this Cosmetic Acu-Point program, I supplemented it with a number of good commercially-available skin care products. The two listed below were the most popular and effective ones used at the Center.

Nu Skin - Skin Care Products

The combination of natural elements used in formulating the Nu Skin line (aloe vera, NaPCA, vitamins A, B, human placental extract, RNA, jojoba, etc.) makes their facial skin care products optimal for skin care. The product line provides unique anti-aging properties that helps skin preserve its inner moisture while retaining its youthful resiliency. The four products used at the Center were the Cleansing Lotion, Ph Balance, Face Activator, and Rejuvenating Cream. For your local distributor call (801) 377-6057.

Clinique - Exfoliation Facial Care System

This three-step, twice-a-day, skin care regimen used at the Center consisted of a gentle Facial Soap that prepared the skin for an exfoliator lotion called Clarifying Lotion. This unique lotion removes the unnecessary layers of dead and dying cells that accumulate every day. The Center used the Dramatically Different Moisturizing Lotion to help nourish the skin. For your local distributor call (212) 572-3800.

CHAPTER ELEVEN

REGAINING CONTROL
30 DAYS OF CHANGE

Eliminating Stress Induced Weight Loss Saboteurs
The Program
Answering General Questions and Concerns
How Many Meals Should I Eat Each Day?
What About Low Blood Sugar?
What About Aids to Digestion?
How Many Calories Should I Eat Each Day?
Do I Need Special Vitamins?
Cooperation is the Key to Success
Eight Keys to Unlock Weight Loss Goals

This chapter contains program calendars to help you eliminate your stress induced weight loss saboteurs. They are meant to give you a visual picture of the consistency of your actions. It is this consistency of your action that will eliminate the stress of weight loss from your life. This chapter also contains the questions and answers that were most frequently asked at The Monterey Wellness Center workshops.

Notes:

Thoughts:

Actions:

Questions:

Reference Pages Found in this Book:

Eliminate Unsuspected Stress Foods (B.R.T.) - pages 73 & 81
Eliminate Food Addictions - pages 99 & 100
Eliminate Toxic Waste Back-Up - page 123
Eliminate Negative Behavior Reactions - page 135
Eliminate Facial Lines and Wrinkles - page 158

CHAPTER   ELEVEN

Regaining Control
30 Days of Change

"A tree that takes both arms to encircle grew from a tiny rootlet.
A many storied building is built by placing one brick upon the next.
A journey of 3,000 miles is begun by a single step."

Lao-Tzu - 6th Century Chinese Philosopher

Eliminating Weight Loss Saboteurs

On the following pages, you will find two progress calendars to help you complete the 30-day program needed to eliminate weight loss saboteurs. Using this organized system, the time needed to make this vital change is minimal.

Prior to eating any food, take a few seconds to monitor that food (using the B.R.T.) to see if it is a fat-producing Unsuspected Stress Food. Anytime you feel a craving for a food that is not part of your dietary regimen, erase the craving with the anti-addiction techniques. Eliminate dieting's toxic waste back-up problems by massaging a few acupressure points each day. In the evening, spend a few moments massaging other specific acupressure points to erase facial lines and wrinkles created by stress and dieting. And finally, while drifting off to a pleasant sleep, you will present your subconscious mind with new visions and mental pictures that will replace old conditioned negative behavior reaction patterns. As simple as these techniques seem, the results are dramatic.

The two progress calendars will help you complete the 30-day program needed to eliminate your weight loss saboteurs.

Putting It All Together

The information found in *Eliminating The Stress of Weight Loss* can be mastered using a 30-day regimen during which the dieter checks off the completion of each daily task. This visual presentation of the dieter's consistency of action will be of great help in keeping the dieter on the program. Each day, for the next 30 days, the dieter will use this program to eliminate their stress induced weight loss saboteurs.

To Achieve and Maintain Weight Loss Goals.

Use Biofeedback to Eliminate Dietary Distress.

Use Biofeedback to Eliminate Unsuspected Stress Foods.

Eliminate Dieting's Toxic Waste Back-Up Problems.

Eliminate Food Addictions Caused by Anxiety and Stress.

Eliminate Negative Behavioral Eating Reactions Caused by Prior Conditioning.

Eliminate Facial Lines and Wrinkles Caused by Rapid Weight Loss or Stress, Anxiety and Tension.

Day - Date	Monitor Your Food Instinct	Erase Food Addic- tions	Remove Negative Behavior	Remove The Toxic Waste	Erase Facial Lines & Wrinkles
1					
2					
3					
4					
5					
6					
7					
8					
9					
10					
11					
12					
13					
14					
15					

Day - Date	Monitor Your Food Instinct	Erase Food Addic- tions	Remove Negative Behavior	Remove The Toxic Waste	Erase Facial Lines & Wrinkles
16					
17					
18					
19					
20					
21					
22					
23					
24					
25					
26					
27					
28					
28					
30					

Answering General Questions and Concerns

How Many Meals Should I Eat Daily?

The best stress-free diet would consist of three or more small meals eaten throughout the course of the day. Eating one large meal a day, like going on a binge, puts an unrealistic stress on the digestive system. The result is incomplete digestion of food which just adds to weight loss problems. Consumption of light meals spread through out the day maintains optimal energy and wellness levels.

What About Hypo-Glycemia?
(Low Blood Sugars)

Eating frequent small meals throughout the day has proven to be an excellent way to maintain blood sugars at a level setting. Un- controlled fluctuations of blood sugar levels leads to a condition called blood sugar intolerance. If this state of mild illness continues, it will lead to "blood sugar stress levels" making the person's physiologic state slightly more uncomfortable. If allowed to continue, this condition leads to hypo-glycemia. Ultimately, if it is not corrected, certain types of diabetes can develop. These disease states are a terrible price to pay for losing a few pounds.

Fluctuating blood sugars can cause extreme fluctuations in emotional states ranging from mild anxiety to severe depression. Also, the ability to think and act clearly is affected. All these symptoms will defeat the best of weight loss goals.

Your awareness of your Food Instinct Signal will allow you to stay slim for the rest of your life.

167

Fluctuating blood sugars cause extreme fluctuations in emotional states ranging from mild anxiety to depression.

What About Aids to Digestion?

When under stress, our body tends to produce an excess amount of digestive acids in the stomach. If the stress continues over a long period of time, this mechanism exhausts and then it cannot produce enough digestive acids.

Initially, if an individual on a restricted diet is under stress, they can develop ulcers because of the over-production of acid. Later, when this mechanism wears down and the stomach does not produce sufficient digestive acid, the body can no longer digest food adequately and large complex molecules enter the digestive system, causing an unsuspected stress-producing mechanism. Taking digestive enzymes or supplemental digestive aids can help break down foods into single component molecules that are easier to digest. A full discussion of digestive aids is beyond the scope of this book. If you need more information on this subject, it is readily available elsewhere.

How Many Calories Each Day?
What Kind of Exercise Should I Do?

For health reasons, I never recommend that a person eat less than 1,200 calories each day. The number of calories you can consume and still lose weight is directly related to your activity level. Use the following chart as a guide. Realizing the amount of exercise it takes to burn off excess calories is probably the best defense against overeating.

Calories Burned Per Hour of Activity *

Lying down or sleeping.................................80
Sitting..100
Driving a car...120
Standing..140
Domestic work...188
Bicycling (5.5 mph)......................................210
Walking (2.5 mph)..210
Gardening..220
Canoeing (2.5 mph)......................................230
Golf 250
Lawn Mowing (power)..................................250
Bowling...270
Lawn Mowing (manual)................................270
Fencing..300
Rowboating (2.5 mph)..................................300
Swimming (0.25 mph)...................................300
Walking (3.75 mph)......................................300
Badminton 350
Horseback riding..350
Volleyball...350
Roller skating...350
Table tennis...360
Ditch digging...400
Ice skating (10 mph).....................................400
Wood chopping 400
Tennis..420
Water skiing...480
Hill climbing (100 ft. / hr)............................490
Downhill skiing..600
Bicycling (13 mph).......................................660
Scull rowing 840
Running (10 mph)...900

Realizing the amount of exercise it takes to burn off excess calories is probably the best defense against overeating.

*In the
1930's
research
revealed
that certain
substances
existed in
nature that
were "vital
for life".*

Do I Need Special Vitamins?

In the 1930's research revealed that certain substances existed in nature that were "vital for life". These chemical substances called "amines", are specific chemical compounds that exist naturally in healthy individuals. Since it was determined that these amines were vital for life, they were given the name "vita-amines". Today we call them vitamins. When you experience a stressful period of life, your body will become depleted of certain minerals (amines). Eating supplemental supplies of these substances replenishes the depleted supply.

If you are living a life of leisure with minimal stress, eating well-balanced meals and are well taken care of, you might not need vitamins. Most individuals today, especially those following weight loss diet programs, need to include a good broad-based multiple vitamin in their diet to maintain their overall health. I also recommend specific nutritional supports for the immune system, the glandular system and general supports for the digestive system. It is not within the scope of this program to make specific recommendations of this sort. I advise serious dieters to consult a professional who can evaluate all the individual health factors involved or to look into one of the numerous excellent nutrition manuals that are available.

Cooperation is the Key to Success *

Poor eating habits developed in childhood can sabotage the best of diet programs. Take the time to develop adult eating patterns that will be beneficial to your weight loss goals.

Plan each menu one day in advance.
Accept the responsibility for your eating habits. Without planning, you give up control of your diet and will always eat incorrectly.

Confirm your planned menu.
Review your menu before eating. Make sure it is what you planned the previous day. This works as a safeguard for those who rationalize that an extra pat of butter or slice of bread "can't hurt".

Restrict eating to a proper area.
"Proper" is carefully defined as the kitchen table, dining room table, or counter in a public eating establishment. This deprives the addictive eater of tasting food as it is cooking, TV snacking or bedtime nibbling.

Complete the food addiction activities.
Whenever you are feeling an urge or craving for a food that is not part of your weight loss diet, complete the anti-addiction procedure to subdue those food cravings. (Chapters Six and Seven). You can also use this procedure to decrease your excess hunger.

Poor eating habits developed in childhood can sabotage the best of diet programs.

Use the B.R.T. to determine if any of the foods you are about to eat will contribute to your weight problem.

Monitor your reaction to all foods before eating them.
Use the B.R.T. to determine if any of the foods you are about to eat are diet saboteurs that will contribute to your weight problem. If they are, do not eat them.

Concentrate on eating.
Reading, talking on the phone, listening to the radio or watching TV are all taboo, even when eating alone. Table conversation is OK.

Leave something on the plate.
The cue for the end of a meal should never be an empty plate. Get rid of the "clean your plate" habit; it is an addictive behavior, often rooted in the guilt-producing programming of childhood.

Be the last one to finish.
Adopt an unhurried eating style. This eliminates the quick gulping and swallowing that puts large amounts of partially digested food into your system. Such foods usually thwart weight loss goals because the molecules are too big for easy digestion. Take longer for each meal and savor the taste of your food.

Eight Keys to Unlock Weight Loss Goals *

Good foods eaten incorrectly are almost as bad as poor foods eaten well. The body must be able to process all food properly if it is to provide you with the proper building blocks and not trigger your Response Pathway Mechanisms.

Key #1 - Never eat when tired. The digestive system does not break down and assimilate food under stress and fatigue.

Key #2 - When feeling tired or fatigued, drink a meal. In a weakened state, the body can assimilate and appreciate liquids much more than solid foods. Try soup or something nourishing made in a blender.

Key #3 - Drink liquids at least 15 minutes before each meal and never with or after meals. Fluids dilute the digestive juices and slow or stop normal digestion.

Key #4 - Chew food completely, moving it back and forth across your tongue. A good general rule is to chew your food until it is liquid. A small amount of food chewed well produces great quantities of natural body fluids. The end result is that you need less food to fill your stomach.

Key #5 - Eat small amounts of food and always quit before you feel full. It depletes the body of energy to have to digest excessive amounts of food.

Good foods eaten incorrectly are almost as bad as poor foods eaten well.

Key #6 - Eat protein early enough to allow proper digestion. Proteins take from four to fourteen hours to digest. Eating them late in the day does not allow the digestive system enough time to complete the digestion before you sleep.

Key #7 - Eat as much of your food in the raw or uncooked state as possible. The enzymes in raw food are the carriers of minerals to the individual cells.

Key #8 - Never eat foods that are too hot or too cold. Extreme temperatures in foods delay and inhibit proper digestion.

* These three sections have been taught at the Center for years. Unfortunately the names of the people who developed them have been lost. I apologize to these authors for using their work without giving them credit but I felt the information was too valuable to omit. Any help in correcting future editions would be greatly appreciated.

ADDENDUM #1

YOUR PERSONAL SELF-TESTING FOOD INSTINCT MONITOR

The Self-Testing Biofeedback Tool
Directions
The Food Instinct Monitor

The Biofeedback Reflex Test learned in Chapters Three and Four enables the dieter to detect their Unsuspected Stress Foods. For those dieters who would like to monitor their Food Instinct Signal without the aid of a friend, Addendum #1 introduces you to the Jamar® Dynamometer. This tool can be found in most doctor's offices. It can also be purchased from any medical supply store. If the dieter is friendly with their physician, there is no medical reason why the dieter could not use the dynamometer for the few minutes that the testing requires. This chapter gives the dieter simple instructions on the use of this monitoring instrument.

This addendum presents two tools that dieters can use for self-testing of their Food Instinct Signal

- The Jamar® Dynamometer (Figure A-1) is available from your local medical supply dealer.

- The Food Instinct Monitor (Figure A-6) is available from the Monterey Wellness Center. Please use the order form on page 223 and 224.

Notes:

Actions:

Questions:

Additional References:

Chapters Three and Four - The Food Instinct Signal
Kinesiology and the Exact-O-Cizer by Dwayne B. Lowell

ADDENDUM #1

Your Personal Self-Testing Food Instinct Monitor

The Self-Testing Biofeedback Tool

The Jamar® Dynamometer is the perfect tool for those individuals who wish to monitor their Food Instinct Signal by themselves - without the aid of a friend. It is a grip monitor that allows physicians to observe variations in grip strength. It is a tool that is readily available anywhere in the country. Most doctors have one in their office plus it is available through medical supply offices. There is no medical reason to prevent its use by the general public.

The use of the dynamometer is fascinating and exciting. If you are on friendly terms with your physician, ask him or her to allow you to use their dynamometer the next time you visit their office. Since most diet programs limit food choices to 20 or 30 different items, the testing should only take ten to fifteen minutes.

When monitoring responses to determine the Unsuspected Stress Foods in your diet, you will notice some foods actually increase your power grip. These foods represent your "Power Foods". Details on the benefits of these foods goes beyond the range of this book. If you want to know more about why these foods are important to you, see my book *Life Empowerment*.

The Jamar® Dynamometer is a fascinatingly simple tool that most doctors have in their office. Ask to use it!!

Figure A.1
Jamar® Dynamometer

Directions for Self-Testing

If your biofeedback response, the Food Instinct, is activated by a food, you will notice a decrease in your ability to squeeze the bars.

Step 1- Determine Baseline Strength
Place the dynamometer in the palm of your hand with the index finger against the finger guide. Squeeze firmly a few times to get the feel of the tool. Move the strength indicator arrow back to zero. Squeeze the dynamometer firmly again. Hold that position for five seconds, release and then read the number pointed to by the indicator arrow. You are looking for the numerical value which you can comfortably hold for the count of five seconds. This will be your baseline reading (Figure A.2). If your biofeedback response, the Food Instinct, is activated by a food, you will notice a decrease in your ability to re-achieve this grip strength reading.

Step 2
Place the foods to be tested in front of you. Remember that the muscles used to monitor any decreases in grip strength will become fatigued. Depending on your health and strength, a maximum of ten to twenty foods is recommended.

Step 3
Take a very small bite of the first food. Chew it a few times and hold it under your tongue. With the food still under your tongue, slowly squeeze the dynamometer again to observe your baseline reading. Note any deviation from your first baseline reading. Is there any change in your ability to maintain your baseline reading for five seconds? Is it harder to do? Is it easier?

Figure A.2
Baseline Position

Did that food make your grip stronger or weaker (Figure A.3)? Did it have any effect at all?

Step 4
Rinse and clear out your mouth with a drink of water.

Step 5
Check your baseline reading again by setting the pointer to zero. Squeeze the monitor and hold comfortably for five seconds to redeter- mine your baseline reading. It is from this reading that you will observe deviations.

Step 6
Test your response to the next food. After-wards, rinse your mouth and repeat Step #3. Continue repeating steps 3 and 4 until you have observed how you are affected by each of the foods tested.

Step 7
Keep a record of all foods that consistently alter your energy levels. If you have one, ask your nutritionist if these foods fall into any specific categories. See if you can determine why these foods are affecting you.

Step 8
Avoid foods that decrease your strength. These foods are your Unsuspected Stress Foods. They are a source of unnecessary stress to your system.

Avoid foods that decrease your strength. These foods are a source of unnecessary stress to your system.

Figure A.3
A Decrease in Strength

179

The Food Instinct Monitor is an inexpensive biofeedback tool that allows you to monitor your own responses by yourself.

The Food Instinct Monitor

The Food Instinct Monitor (Figure A.6) is an inexpensive ($25.00) biofeedback tool that is available only through The Monterey Wellness Center (pages 223 and 225). This simplified version of the Jamar Dynamometer also allows you to observe your Food Instinct Reflex. The simplicity and economy of this tool is based on the fact that you can observe a sliding ring to determine decreases in grip strength in response to Unsuspected Stress Foods.

Once you determine your baseline numerical reading (Figure A.4), you can use this tool to observe fluctuations in your grip strength. It is this fluctuation - a decrease in your measurable grip strength (Figure A.5) that will allow you to recognize foods that will enter your system as complex molecules.

On the grip scale, the distance from one number to the next represents approximately eight pounds of grip strength. Therefore, if your ability to re-achieve baseline reading decreases from six to four (Figures A.4 and A.5), this represents a decrease of approximately 16 pounds of grip strength.

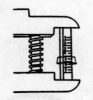

Figure A.4
Baseline Strength

Figure A.5
Decrease in Strength

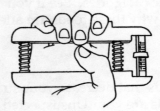

Figure A.6
The Food Instinct Monitor

ADDENDUM #2

THE CLINICALLY
REPRODUCIBLE PROOF

A Valid Observation
The Mouth/Food Relationship
The Equipment
Premise
The Experiment
The Consistency of Test Foods
The Variation Factor
The Results

Addendum #2 presents a clinically reproducible experiment that demonstrates the muscle reflex response caused by stress foods. Dr. Shain and the staff of The Monterey Wellness Center welcome feedback from individuals or institutions who are in a position to reproduce this experiment.

Notes:

Thoughts:

Actions:

Questions:

ADDENDUM #2

The Clinically Reproducible Observation

Abstract:
The introduction of a non-specific food into the digestive pathways of the human body can produce an instantaneous and recordable fluctuation in the strength of body muscles.

A Valid Observation

Since the beginning of recorded history, man has been observing and recording observations about the world around him. His hope has been that once understood, the information could then be used to advance his life along the path of self-control and self-determination. Once an observation is made, a researcher must determine if the same observation can be consistently reproduced and observed by other researchers. Once the consistency of the observation has been satisfied, then the scientist studies and explains why that observation is consistent. Once the "why" is known, the information can be used to advance humankind in a multitude of directions.

The mechanisms detailed in this book, involving the relationship between food and the human body's reactions to food, warrant a great effort by the scientific

Since the beginning of recorded history, man has been observing and recording observations about the world around him.

The following clinically reproducible experiment is provided to help the scientific community accept the challenge of these observations.

community to help explain the "whys" that are at work in this process. The information in this book are the "whys" I have observed over the past 14 years and the ways in which I have applied this information to help guide my patients to a happier and healthier life.

The following clinically reproducible experiment is provided to help the scientific community accept the challenge of this observation. It can be completed anywhere in the world by any individual following the included set of written instructions.

So that the results of the experiment would not be subjective (i.e. asking the people being tested what they think happened), the experimental model explained in the section uses computerized testing and research equipment that is available in many hospitals across the United States. Thus, the accumulated data can be processed in exactly the same manner wherever the experiment is being conducted.

The other test factor that must be consistent is the food. It is essential that it be fresh food prepared within an hour or two of the testing. It must be prepared in the same manner and have the same consistency of taste, texture and smell. This test uses food from a popular chain restaurant that meets these requirements.

The Mouth/Food Relationship

The most common questions about the monitoring procedure concern the speed of the observation. People want to know why they do not have to wait for the food to enter the digestive system. They ask if the digested molecules have to enter the brain to effect a change. Also, they want to know why the muscle strength goes back to baseline so quickly when the mouth is rinsed with water.

The act of chewing, tasting and smelling food all work instantaneously to let your brain know what you have decided to ingest. As soon as your brain recognizes that food (i.e., you become consciously aware of the "taste" of that food), the central nervous system (the brain) signals the correct digestive organs needed to digest the food. The biofeedback signal we are monitoring comes from this central nervous system signal, not from the digestive mechanisms. Therefore, the response is only there for the first few moments after tasting the food in your mouth. Also, after you rinse your mouth and no longer taste the food, there is no need for your central nervous system to continue sending the signal, thus the response is no longer present.

The biofeedback signal comes from this central nervous system signal, not the digestive mechanisms.

The Equipment

A computerized exercise, testing and research system developed by The Chattecx Corporation of Chattanooga, Tennessee was used to monitor and record any variations in each "three-trial" test. The Kin-Com unit is a hydraulically-powered, computer-controlled, muscle resistance monitoring system. Its data capture system collects data on force, velocity and range of motion. This data is used to calculate the power, torque and work units which are used to generate reports.

Premise

If the introduction of a food morsel into the mouth of the person being tested has any effect upon the muscle strength of the muscle being monitored, then the measurement would be different from any before and after trials in which there was no food in the subject's mouth. Refer to Examples #1, #2, and #3 for possible variations of any three trial tests.

The Experiment
(Ten tests of three trials each)

Determine the Baseline "In the Clear"

The power output of the right biceps muscle, measured in watts, is recorded in three separate trials over a fifteen-second span. Each time the strength is monitored with no food in the test subject's mouth, it is referred to as recording the muscle strength "In the Clear". A computer-

Strength of Tested Muscle
0......10......20......

Trial 1
Trial 2
Trial 3

Example #1
All trials show output
of equal strength.

generated chart is printed showing the maximum strength generated by this one muscle on each trial. For consistency of comparison, the individual is asked to rinse her mouth with a drink of water prior to each trial.

Test #1 through #10 - The Food Instinct

The power output of the right biceps muscle, measured in watts, is recorded in three separate trials over a fifteen-second span. A computer-generated chart is printed showing the maximum strength generated by this one muscle on each trial. Trial #1 is recorded "In the Clear." Immediately prior to Trial #2, the individual chews and holds a given food morsel under her tongue. Trial #2 is run and recorded. Then the individual rinses her mouth so she no longer tastes the food. Then Trial #3 is performed "In the Clear". This procedure is repeated until all ten foods are monitored.

End- Confirm Original Baseline "In the Clear"

The power output of the right biceps muscle, measured in watts, is recorded in three separate trials over a fifteen-second span. A computer-generated chart is printed showing the maximum strength generated by this one muscle on each trial. For consistency of comparison, the individual is asked to rinse her mouth with a drink of water prior to each trial. Each time the strength is monitored in this manner, it is referred to as recording the muscle strength "In the Clear".

Strength of Tested Muscle
0.....10..... 20....

Trial 1
Trial 2
Trial 3

Example #2
A decrease in strength is registered in Trial #2 with food in mouth.

Strength of Tested Muscle
0......10......20.....

Trial 1
Trial 2
Trial 3

Example #3
An increase in strength is registered in Trial #2 with food in mouth.

The foods used in this experimental model should be nutritional, healthful foods that would be considered excellent choices for any weight loss regimen.

The Consistency of Test Foods

The foods used in this experimental model should be nutritional, healthful foods that would be considered excellent choices for any weight loss regimen. They should be fresh and of the same consistent quality no matter where they are purchased. Do not use "junk foods" or other "chemically processed" foods for they are food choices dieters should avoid.

To meet these requirements on a national scale, I chose an easily accessible, fast food restaurant that has demonstrated consistent quality standards. Over the years this organization has quietly supplied dieters with fresh, consistently healthful, low-calorie food choices.

Purchase foods for this experiment from your local "Jack-In-The-Box" restaurant a maximum of a few hours prior to running the tests. I selected foods from the following menu items: Chicken Fajita Pita (292 calories); Scrambled Egg Platter; Chef Salad (325 calories) and a plain hamburger with no dressing.

1) Grilled chicken 2) Pita bread
3) Scrambled eggs
4) Grilled breakfast potatoes
5) French fried potatoes
6) Ham 7) Lettuce 8) Tomato
9) Ground beef (from plain hamburger)
10) Iced Tea

The Variation Factor

In any test procedure, there is a plus or minus factor which must be used to temper the results. This test model measures the strength of the biceps muscle in the unit of power called "watts". The error factor for dynamometer strength comparisons is plus or minus ten. Therefore, in order for any variation in test Trial #2 to be significant, the measurement must be at least ten watts above or below the average reading of Trial #1 and Trial #3.

The Results

The graph results of testing one person are shown on page 190. Notice that of the ten foods tested, three foods for this person showed a significant variation in Trial #2.

Test #1 showed an 18 point decrease in strength.
Test #7 showed a 16 point decrease in strength.
Test #10 showed a 19 point decrease in strength.

Retesting this individual on subsequent testing days showed the findings to be consistent. Since each dieter tested will differ based on their health, lifestyle, stress factors and heredity, how they respond to a specific food also varies. Some people found that none of these foods triggered their Food Instinct. Some discovered only one food that did. Some people discovered many foods that triggered their Food Instinct. The results for each individual tested proved to be consistent during subsequent testing.

Since each dieter tested will differ based on their health, lifestyle, stress factors and heredity, how they respond to a specific food also varies.

Average Strength of Tested Muscle

0......10......20......30......40......50

Of the ten foods tested, three foods for this person showed a significant variation in Trial #2.

Test #1 Trial#1	
Trial #2	
Trial #3	
Test #2 Trial#1	
Trial #2	
Trial #3	
Test #3 Trial#1	
Trial #2	
Trial #3	
Test #4 Trial#1	
Trial #2	
Trial #3	
Test #5 Trial#1	
Trial #2	
Trial #3	
Test #6 Trial#1	
Trial #2	
Trial #3	
Test #7 Trial#1	
Trial #2	
Trial #3	
Test #8 Trial#1	
Trial #2	
Trial #3	
Test #9 Trial#1	
Trial #2	
Trial #3	
Test#10Trial#1	
Trial #2	
Trial #3	

AFTERWORD

Eliminating The Stress of Weight Loss, started as a small class at The Monterey Wellness Center in 1980. We watched it grow and mature just as a child. And just as it is with a child, each stage was exciting and special.

Growth though, is a never-ending process. It continues throughout all aspects of life. We understand better as we gain more information. We then reshape and update.

I would love to hear from you - the reader of *Eliminating The Stress of Weight Loss.*

It is you alone who can now provide me with more understanding - by providing me with more information. Share your successes with me. Tell me what chapters were particularly helpful. Give me your suggestions for improvement.

I assure you that I will answer your letters.

Eliminating The Stress of Weight Loss will give you a unique experience in mastering control of your life as it reacts to your daily stressors.

Readers will be pleased to discover that this experience also revitalizes and recharges the body, mind and spirit.

It is my greatest hope that this program will help you to also discover the ease by which you can control all other directions and events of your life.

Best of luck to you!

APPENDICES

5,000 SUPPORT PERSONNEL WHO CAN HELP YOU UNDERSTAND THIS PROGRAM

Help from Your Neighbors and Friends

Since 1974 over two million people have studied from a marvelous book called *Touch For Health*. This program teaches the use of acupressure, touch and massage to promote better health. There are approximately 3,000 instructors qualified to teach this program in the United States. Any of these instructors and most of the individuals who have taken their classes can be of assistance in understanding *Eliminating The Stress of Weight Loss*. For an instructor in your area, or for the book, *Touch For Health*, contact Touch For Health at 1200 N. Lake Ave., Pasadena, California 91104 or call (818) 798-7893.

Help From Physicians Familiar With This Work

The International College of Applied Kinesiology (I.C.A.K.), founded in 1974, has an active membership of over 1,000 doctors nationally who are involved in research and studies of the principles outlined in *Eliminating The Stress of Weight Loss*. In addition, the I.C.A.K. maintains lists of other physicians (not involved in research) who have studied and have a working knowledge of the subject. If you, your physician, dietician or counselor need more professional information, concerning The Art and Science of Applied Kinesiology or a practicing doctor in your area, contact the I.C.A.K. at P.O. Box 25276, Shawnee Mission, Kansas 66225 or call (913) 648-2828.

Help From Physicians Familiar With This Work

Dr. Scott Walker, of Encinitas, California, has developed a body of work called **The Neuro Emotional Complex**. Thousands of doctors have studied this work. They then can help their patients by neutralizing the effects of Past Emotional Trauma. It is Past Emotional Traumas that sabotage both health and well-being. These doctors all have a working knowledge of the information presented in this book. For a doctor in your area, call or write; N.E.T., 500 Second Street, Encinitas, California 92924 or call (619) 944-1030.

Books and Other Programs

T.H. Enterprises, 1200 N. Lake Ave., Pasadena, California 91104 or call (818) 798-7893. An excellent resource for books and other programs mentioned in *Eliminating The Stress of Weight Loss*. Contact them for a copy of their marvelous catalog.

Books for Physicians

Systems D.C., 275 W. Abriendo Ave., Pueblo, Colorado 81004 or call (719) 543-3617. An excellent resource for physicians wishing to learn more about the subject of Applied Kinesiology. Books, educational materials and program tapes for the professional.

Rate Your Present Diet Program

Consumer Guide has published a book called, *Rating the Diets* by Theodore Berland. Published in 1984, this book evaluates, in a professional unbiased manner, 108 current diet programs available in book form. Although it is slightly dated, it is highly recommended.

26 ADDITIONAL INFORMATION SOURCES TO ANSWER YOUR DIET QUESTIONS AND CONCERNS

American Association of Nutritional Consultants (AANC) - 1641 E. Sunset Rd., B117, Las Vegas, NV 89119 Founded: 1980. Members: 5000. Staff: 5. Professional nutritional consultants. (702) 361-1132

American Chiropractic Association Council on Nutrition - Dr. Alan Pressman 7 East 9th St., New York, N. Y. 10003 Founded: 1975. Members: 2,000. Chiropractic Physicians qualified to treat nutritional and metabolic disorders. (212) 228-5600

American College of Nutrition (ACN) - 722 Robert E. Lee Drive, Wilmington, N.C. 28412-0927 Founded: 1959. Members: 900. Provides education on clinical and experimental developments in the field of nutrition.

American Council of Applied Clinical Nutrition (ACACN) - PO Box 504, Florissant, MO 63032 Founded: 1974. Members: 500. Staff: 6. Organization of clinical nutrition specialists.

American Dietetic Association (ADA) - 216 W. Jackson Blvd., Suite 800, Chicago, IL 60606 Founded: 1917. Members: 60,000. Staff: 120. Food and nutrition professionals.

American Society for Clinical Nutrition (ASCN) - 9650 Rockville Pike, Bethesda, MD 20814 Founded: 1959. Members: 1025. Physicians and scientists actively engaged in clinical nutrition research.

Buxom Belles, International (BBI) - 27856 Palomino Dr., Warren, MI 48093 Founded: 1956. Members: 1600. An organization that provides support and assistance to overweight men and women on a group-therapy basis.

Clinical Kinesiology Organization for Research and Education - Dr. Gary N. Klepper, D.C. 1440 28th Street, Boulder, CO 80303 (303) 449-7388. Research organization of physicians engaged in research and education concerning Clinical Kinesiology observations.

Cooking for Survival Consciousness (CSC) - Box 26762, Elkins Park, PA 19117 Founded: 1977. An educational and scientific membership foundation dedicated to disseminating information concerning the relationship between good health, food and cooking.

Feingold Association of the United States (FAUS) - PO Box 6550, Alexandria, VA 22306 Founded: 1976. Members: 30,000. (703) 768-3287. Programs designed to eliminate the dietary reasons for the hyperactivity of children and the chemical-sensitivities of adults.

Food and Nutrition Board (FNB) - Institute of Medicine. 2101 Constitution Ave., N.W., Washington, DC 20418 Founded: 1940. Members: 16. Staff: 15. (202) 334-1737. A professional organization that evaluates the relationship between food consumption, nutritional status, and public health.

International Academy of Nutrition and Preventive Medicine (IANPM) - PO Box 18433 Ashville, N.C. 28814 (704) 258-3243 Founded: 1971. Members: 700. A professional group that supports and encourages the use of nutrition and preventive medicine as part of our health care system.

International College of Applied Kinesiology - P.O. Box 25276, Shawnee Mission, Kansas 66225 (913) 648-2828. Founded: 1973: Members: 1,000. Physicians qualified in treating dysfunction using as part of their therapeutic regimen the science of Applied Kinesiology.

Institute of Behavioral Kinesiology - Physicians qualified to treat behavioral and emotional problems using as part of their therapeutic regimen the science of Applied Kinesiology: Contact The Monterey Wellness Center for details.

Lean Line (LL) - 151 New World Way, South Plainfield, NJ 07080 Founded: 1968. An optimum health/wellness program of nutritional, psychological / image-building techniques.

Nathan Pritikin Research Foundation (NPRF) - 1910 Ocean Front Walk, Santa Monica, CA 90405 (213) 450-5433. RA professional organization researching the relationship between diet, lifestyle and chronic disease.

National Association to Advance Fat Acceptance (NAAFA) - PO Box 188620, Sacramento, CA 95818 Founded: 1969. Members: 3000. An organization dedicated to improving the quality of life for fat people in ways other than dieting.

Natural Food Associates (NFA) - PO Box 210, Atlanta, TX 75551 Founded: 1952. Members: 10,000. Staff: 3. State Groups: 29. Local Groups: 100. Professionals and consumers interested in organic farming, natural foods, preventive medicine and health education.

Nutrition Education Association (NEA) - PO Box 20301, 3647 Glen Haven, Houston, TX 77225 Founded: 1977. Subscribers: 25,000. Staff: 5. Professionals and other individuals interested in educating the public on the relationship between good health and nutrition.

Nutrition for Optimal Health Association (NOH) - PO Box 380, Winnetka, IL 60093 (708) 835-5030 Founded: 1972. Members: 500. Nonprofit educational organization that promotes knowledge and use of correct nutrition to achieve and maintain optimal health.

O-Anon General Service Office - PO Box 4305, San Pedro, CA 90731 Founded: 1985. Regional Groups: 60. An organization that provides information and support to families and friends of compulsive overeaters.

Overeaters Anonymous (Compulsive Overeating) (OA) - PO Box 92870, Los Angeles, CA 90009 (213) 542-8363 Founded: 1960. Staff: 15. Worldwide Groups: 10,500. Men and women who meet to share their experiences, strength, and hope in order to arrest the disease of compulsive overeating.

TOPS Club (Take Off Pounds Sensibly) - PO Box 07360, 4575 S. Fifth St., Milwaukee, WI 53207 Founded: 1948. Members: 305,650. International Chapters: 11,650. Weight control self-help association using group therapy, competition, and recognition to help members lose weight. (800) 932-8677.

Touch For Health Foundation - 1200 N. Lake Ave., Pasadena, CA 91104 (818) 794-7893 Founded: 1974. Members: 13,000, Staff: 5. A non-profit organization devoted to teaching the benefits of accupressure, touch and massage to achieve greater health.

Weight Watchers International (WWI) - Jericho Atrium, 500 N. Broadway, Jericho, NY 11753 Founded: 1963. An organization that provides support and guidance for people wishing to control weight.

RECOMMENDED READING / BIBLIOGRAPHY

The ISBN number is a code number used to locate a specific book.

ABC of Nutrition by A. Stewart Truswell (ISBN 0-0-7279-0117-6), Taylor and Francis, 1986

Acupressure: The Ancient Art of Healing and How It Works Scientifically by Dr. Felix Mann (ISBN 0-8446-4583-4), Random, 1973

Acupressure Weight Loss by Michael Reed Gach, Acupressure Institute, Berkeley, Ca.

Acu-Facelift by Michael Reed Gach, Acupressure Inst., Berkeley, Ca.

Addictive Behaviors: The Treatment of Alcoholism, by William R. Miller (ISBN 0-08-030837-6), Pergamon, 1983

America and Its Food: A History of Food Habits in the United States by Richard O. Cummings (0-405-02445-2), Ayer Co Pub, 1970

Applied Kinesiology Synopsis by David Walther, D.C. Systems DC, Co, 1986

As A Man Thinketh by James Allen (ISBN 0-399-12829-8), Gros. & Dun., Putnam Pub., 1983

Aspects of Food Production, Consumption and Energy Values edited by GH Bourne (ISBN 3-8055-4993-8), S. Karger, 1989

At-A-Glance Nutrition Counter by Patricia Hausman (ISBN 0-345-31183-3), Ballantine, 1984

Behavioral Kinesiology and the Autonomic Nervous System by John Diamond, M.D. (ISBN 0-3960849-3), Archaeus Press, 1978

Better Homes and Gardens Calorie-Counter's Cook Book by Better Homes and Gardens Editors (0-696-00835-1), BH&G, 1983

Body Time, Physiologic Rhythms and Social Stress by Gay Gaer Luce (ISBN 0-48622586-0), Dover, New York, 1971

Bodybuilder's Nutrition Book, The by Franco Columbo and Lydia Fragomeni (ISBN 0-8092-5457-3), Contemp Bks, 1985

Body Quantum by Fred Alan Wolf (0-02094080-7) Macmillan 1986

Breakthrough for Dyslexia & Learning Disabilities by Dr. Carl Ferreri and Dr. Richard Wainwright, New York, 1984

California Nutrition Book: Food for the Ninties - University of California Faculty and American Health Magazine by Paul Saltman et al (ISBN 0-316-76964-9), Little, 1987

Choose To Be Well Diet Cookbook by Beatrice Wittels (ISBN 0-9610780-0-6), CSC Pub. 1989

Coaches Guide to Nutrition and Weight Control by Patricia Eisenman and Dennis Johnson (0-931250-25-0), Human Kinetics, 1982

Complete Book of Spices: Their Medical, Nutritional and Cooking Uses by John Heinerman (ISBN 0-87983-347-5), Keats, 1983

Controversies in Clinical Nutrition edited by John J. Cunningham (ISBN 0-89313-021-4), GF Stickley, Co, 1980

Creative Visualization by Shakti Gawain (ISBN 0-553-24147-8), New World Library, 1978

CRC Handbook of Nutritive Value of Processed Food edited by M. Rechcigl, Jr. (ISBN '0-8493-3951-0,TX551), CRC Pr., 1982

Culinary Kid: A Nutrition Guide and Cookbook for Parents and Kids by St. Joseph Hospital (0-9616857-1-9), St. Joseph Hospital, 1986

Diet and Nutrition by R. Ballentine (ISBN 0-89389-048-0), Himalayan Pub. 1984

Diet For A Small Planet: 10th Aniversity Edition by Frances M. Lappe (ISBN 0-345-32120-0), Ballantine, 1985

Dynamic Laws of Healing, The by Catherine Ponder Devorss and Co., CA, 1966

Eat Better, Live Better: A Common Sense Guide to Nutrition and Good Health by Reader's Digest Editors (ISBN 0-89577-141-1), Random (RD Assn.), 1985

Eating and Weight Disorders: Advances in Treatment and Research edited by Rich L. Goodstein (0-8261-3830-6), Springer Pub., 1983

Energy Intake and Activity by Ernesto Pollitt and Peggy Amante(ISBN 0-8451-1610-X), ARLiss, 1984

Energy: Nutrition of Women edited by G.H. Bourne (ISBN 0-3-8055-4664-5), S. Karger, 1987

Evaluation and Management of Eating Disorders: Anorexia, Bulemia and Obesity edited by Kristin L. Clark, et al (ISBN 0-87322-911-8), Human Kinetics, 1988

Excessive Appetites: A Psychological View of Addiction by J. Orford (ISBN 0-471-10301-2), Wiley, 1985

Fat Distribution During Growth and Later Health Outcomes edited by Claude Bouchard and Francis E. Johnston (ISBN 0-471-61252-9), Wiley, 1988

Food Combining For Health by Doris Grant and Jean Joice (ISBN 0-7225-0882-4), I.T., 1987

Food Power: A Doctor's Guide to Common Sense Nutrition by E. Earle Arnow (ISBN 0-911012-37-0), Nelson-Hall, 1972

Food and Weight Control: Expanded Edition by Brent Q. Hafen (ISBN 0-697-06820-X), Longwood Div (Wm. C. Brown), 1981

Food and Nutrition by Prevention Magazine Editors and Nancy Nugent (ISBN 0-87857-464-6), Rodale Pr. Inc., 1984

Food Allergies by Neil S. Orenstein and Sarah L. Bingham (ISBN 0-317-59799-X), Put. 1988

Food and Fitness by Linda Carlson (ISBN 0-89586-622-6,HP Bks), Price Stern, 1988

Food and Nutrition: Customs and Culture by Paul Fieldhouse (ISBN 0-7099-1079-7), Routledge Chapman and Hall, 1985

Food Composition and Nutritional Tables by S.W. Souci and W. Fachmann (ISBN 3-8047-7), Wissenschaftliche W. Germany, 1986,87

Freedom From Chronic Disease by Arthur L. Kaslow and Richard B. Miles (ISBN 087477318-0), J.P. Tarcher Inc., 1979

Healing Mind, The by Irving Oyle, (ISBN 0-912310-80-4), Celestial, 1974

How To Live To Be One Hundred Or More by George Burns (ISBN 0-39912787-9), Putnam Pub Group, 1983

Inner Bridges by Fritz Fredrick Smith, (ISBN 0-89334-086-3), Hum, 1986

Joy's Way by Brugh Joy, M.D., (ISBN 0-87477-085-8), J.P. Tarcher Inc. 1979

Know Your Nutrition by Linda Clark, (ISBN 0-87983-401-3) Keats, 1984

Let's Eat Right to Keep Fit by Adelle Davis (ISBN 0-451-15550-5, Sig), NAL, 1970

Low Blood Sugar Cookbook, The by Patricia Krimmel and Edward Krimmel (ISBN 0-916503-01-1), Franklin Bryn Mawr, 1986

Medical Applications of Clinical Nutrition by Jeffrey Bland (ISBN 0-87983-327-0), Keats, 1983

Mental Health Through Nutrition by Tom R. Blaine (ISBN 0-8065-0424-2), Carol Pub, 1974

Mind as Healer, Mind as Slayer by Kenneth R. Pelletier (ISBN 038528646-5), Dell Publishing, New York, 1977

MRT (Muscle Response Test) by Dr. Walter Fishman and Dr. Mark Grinims (ISBN 399-90011-x), R. Marek, 1979

New Beauty - The Acupressure Facelift by Lindsay Wagner and Robert M. Klein (ISBN 0-13-536806-5), Prentice Hall, 1986

New Life Through Nutrition by Dr. Sheldon Deal, (602-323-7133)), N.L. Press, Tu. Az.1974

New Life Through Natural Methods by Dr. Sheldon Deal, (602-323-7133), N.L. Press 1979

New Mansions for New Men by Dane Rudhyar (ISBN 08973002-9), El Camino Press, 1978

Nutrition and Athletic Performance edited by William Haskell et. al.(ISBN 0-915950-56-1), Bull Pub., 1982

Nutrition and Physical Degeneration by Weston A. Price (ISBN 0-916764-00-1), Price-Pottenger, 1977

Nutrition and Health Encyclopedia, The by David F. Tver and Percy Russell (ISBN 0-442-24859-8), Van Nos Reinhold, 1981

Nutrition, Disease, Resistance and Immune Function by Watson (ISBN 0-8247-7167-2), Dekker, 1984

Nutrition and Diet Therapy by Carolynn E. Townsend (ISBN 0-8273-2422-7), Donn. 1985

Nutrition for Vegetarians by Agatha Thrash and Calvin Thrash (ISBN 0-942658-03-5), Thrash Pub., 1988

Nutrition and Mental Illness: An Orthomolecular Approach to Balancing Body Chemistry by Carl C. Pfeiffer (ISBN 0-89281-22605), Healing Arts Pr. (Inner Tradit), 1988

Nutrition and Diet Therapy: Self-Instruction Modules by Peggy and Hui Stanfield (ISBN 0-86720-358-7), Jones and Bartlett, 1986

Nutrition and Athletic Performance by Ellington Darden (ISBN 0-87095-058-4, Athletic), Bldn West Bks, 1975

Nutrition and Your Body by Benjamin Colimore and Sarah Colimore (ISBN 0-8095-6302-9), Borgo Pr, 1987

Nutrition and Vitamin Therapy by Michael Lesser, M.D. Grove Press, Inc., Ca, 1980

Nutritional Value of Cereal Products, Beans and Starches edited by G.H. Bourne (ISBN 0-3-8055-4992-X), S. Karger, 1989

Nutritional Self-Defense: A Nutritional Handbook by Lily Splane (ISBN 0-945962-00-2), Anaphase II, 1988

Nutritional Guidelines for Correcting Behavior by Barbara Reed (ISBN 0-686-83743-6), Natural Pr., 1984

Obesity, Weight Loss and Eating Disorders by Aveline Kushi and Helaine Honig (ISBN 0-87 040-642-6), Japan Pub. USA, 1987

Obesity and Anorexia: A Question of Shape by Peter Dally and Joan Gomez (ISBN 0-571-11472-5), Faber and Faber, 1980

Opportunities in Nutritional Careers by Carol C. Caldwell (ISBN 0-8442-6173-4),Textbk, 1987

Overeater's Guide to Will Power, Vol. 1(ISBN 1-55677-095-2), VHI Library, 1988

Psychology of Consciousness, The by Robert E. Ornstein (ISBN 0-15-573082-7), Harbrace 1975

Physiology, Social Nutrition and Nutritional Education (ISBN 3-8055-3048-X), S Karger, 1981

Physiopathology and Treatment of Functional Disorders, The by George Whatmore, M.D. (ISBN 080890851-0), Grune and Stratton, Inc., New York, 1981

Pritikin Program for Diet and Exercise, The by Nathan Pritikin (ISBN 0553256808-4), New York, Bantam Books, 1980

Procedures for Testing of International Food Additives to Establish Their Safety for Use (Nutrition Meetings Reports:No 17) (ISBN 0-92-5-1201882-7), UNIPUB., 1958

Psycho-Nutrition (Health, Nutrition and Well-Being Books) (ISBN 0-399-50789-2,G&D), Putnam Pub. Group, 1983

Psychology of Successful Weight Control by Mary C. Tyson and Robert Tyson (ISBN 0-88229-103-3), Nelson-Hall, 1974

Rating the Diets by Consumer Guide by Theodore Berland (ISBN 0-451-14291-8), Signet, 1984

Recent Advances in Therapeutic Diets compiled by Clinical Staff, Dietary Dept., University of Iowa Hospitals and Clinics (ISBN 0-8138-1064-7), Iowa St U Pr, 1989

Reflections in the Light by Shakti Gawain (ISBN 0-931432-13-8), New World Library, 1988

Sex, Nutrition and You by Gordon S. Tessler (ISBN 0-0-932213-61-8), Better Health, 1986

Stress of Life by Hans Selye, M.D. (ISBN 0-07056212-1), McGraw Hill, New York, 1976

Stress and the Art of Biofeedback by Barbara B. Brown, Ph.D (ISBN 9-84639-857-0), Signet, 1977

Stress, Sanity and Survival by Robert Woolfolk (ISBN 04514848-7), Signet, New York, 1978

Story of Weight Watchers, The by Jean Nidetch (ISBN 04530101-8), New York, New American Library, 1972

Surgeon General's Report on Nutrition and Health by The United States Department of Health and Human Services Staff (ISBN 0-446-39061-5), Warner Bks, 1989

Tao of Balanced Diet: Secrets of a Thin and Healthy Body, The by Stephen T. Chang (ISBN 0-942196-05-8), Tao Pub, 1987

Tao of Health by Michael Blate (ISBN 019687817-1), Falkynor Books, Fl, 1978

Theory of Applied Electro-Acupuncture, The by Thomas Wing, D.C. Seedo Press, Ca, 1974

Thirty-Five Plus Diet for Women: Kaiser Permanente Clinic's Breakthrough Metabolism Diet, The by Jean Spodnik and Barbara Gibbons, (ISBN 0-06-015718-6), Har-Row, 1987

Touch for Health by John F. Thie, D.C. and Mary Marks, Devorss and Co., CA, 1973

Vest Pocket Calories Counter (0-385-00174-6), Doubleday, 1984

Wellness Workbook Regina Ryan and Dr. John Travis (ISBN 0-89815-032-9), Ten Speed Press, 1984

Weight Loss for the Middle-Aged Athlete by B. Vandre (ISBN 0-942223-00-4), Van Press, 1987

Weight, Sex and Marriage by Richard B. Stuart (ISBN 0-671-67008-5), S&S, 1989

Weight Control: A Guide for Counselors and Therapists edited by Aaron M. Atlschul (ISBN 0-275-92697-4,C2697), Praeger, 1987

Weight Watchers Food Plan Diet Cookbook by Jean Nidetch (ISBN 045301010-5) New York, New American Library, 1984

You Are What You Eat by Victor H. Lindlahr (ISBN 0-89370-604-3), Borgo Pr, 1971

You Can Heal Your Life by Louise Hay (ISBN 0-942494-94-6), Hay House, 1984

Your Body Doesn't Lie by Dr. John Diamond (ISBN 044634278-5), Warner Bks, 1979

Your Personalized Health Profile: Choosing the Diet That's Right for You by Myron Winick (ISBN 0-688-05114-6), Morrow, 1985

RECOMMENDED WEIGHT LOSS & FITNESS SPAS
Detailed information about these and other spas can be found in,
The Best Spas by Dr. Theodore B. Van Itallie, Harper & Row, 1988

U.S.A. West Coast

The Ashram Health Retreat: P.O. Box 8009, Calabasas, California 91372, (818) 888-0232 Est. 1974

Cal-a-Vie - The Ultimate Spa: 2249 Somerset Road, Vista, California 92084, (619) 945-2055 Est. 1986

Golden Door: P.O. Box 2548, Escondido, California 92033, (800) 443-7565 Est. 1959

La Costa Hotel and Spa: Costa del Mar Road, Carlsbad, California 92009, (619) 438-9111 Est. 1965

The Oaks at Ojai: 122 East Ojai Ave., Ojai, California 93023, (805) 646-5573 Est. 1970

The Palms at Palm Springs: 572 N. Indian Ave., Palm Springs, California 92262, (619) 325-1111 Est. 1979

Sonoma Mission Inn & Spa: P.O. Box 1447, Sonoma, California 95476, (707) 938-9000

U.S.A. West

Canyon Ranch: 8600 East Rockcliff Road, Tucson, Arizona 85715, (800) 726-9900 Est. 1979

The Cliff Lodge and Spa: Snowbird, Utah 84092, (801) 742-2222 Est. 1987

The Greenhouse: P.O. Box 1144, Arlington, Texas 76004, (817) 640-4000 Est. 1965

Loew's Ventana Canyon Resort: 7000 North Resort Drive, Tucson, Az. 85715, (602) 299-2020 Est. 1987

The Phoenix Fitness Resort: 111 North Post Lane, Houston, Texas 77024, (713) 680-1601 Est. 1980

Rocky Mountain Wellness Spa and Institute: P.O. Box 777, Steamboat Springs, Colorado 80477, (303) 879-7772 Est. 1982

U.S.A. Midwest

Bluegrass Spa: 901 Galloway Road, Stamping Ground, Kentucky 40379, (502) 535-6261 Est. 1985

The Heartland Health and Fitness Retreat: Rural Route #1, Box 181, Gilman, Illinois 60938, (815) 683-2182 Est. 1983

The Kerr House: P.O. Box 363, 17777 Beaver Street, Grand Rapids, Ohio 43522, (419) 832-1733 Est. 1980

The Wooden Door: 628 Mulberry Ct., Milwaukee, WI 53217, (800)-800-7606 Est. 1979

U.S.A. Southeast

Bonaventure Resort and Spa: 250 Racquet Club Road, Fort Lauderdale, Fl. 33326, (305) 389-3300 Est. 1982

Doral Saturnia International Spa Resort: 8755 Northwest 36th Street, Miami, Florida 33178, (305) 593-6030 Est. 1987

Palm-Aire Spa Resort & Country Club: 2501 Palm-Aire Drive North, Pompano Beach, Florida 33069, (305) 972-3300 Est. 1971

Safety Harbor Spa & Fitness Center: 105 North Bayshore Drive, Safety Harbor, Florida 34695, (813) 726-1161 Est. 1926

The Spa at Turnberry Isle: 19735 Turnberry Way, Miami, Florida 33180, (305) 327-7028 Est. 1980

U.S.A. Northeast

Canyon Ranch in the Berkshires: Bellefontaine, Kemble Street, Lenox, Massachusetts 01240, (800) 726-9900 Est. 1989

Deerfield Manor: R.D. #1 (Route 402), East Stroudsburg, Pennsylvania 18301, (717) 223-0160 Est. 1981

The International Health and Beauty Spa at Gurney's Inn: Montauk, Long Island, New York 11954, (516) 668-2345 Est. 1979

Kripalu Center for Yoga and Health: P.O. Box 793, Lenox, Massachusettes 01240, (413) 637-3280 Est. 1971

New Life Fitness Vacations: Inn of the Six Mountains, P.O. Box 395, Killington, Vermont 05751, (802) 422-4302 Est. 1978

England

Champneys at Tring: Tring, Hertfordshire, HP23 6HY, England (011) 44 (4427) 3351, Est 1925

Forest Mere: Liphook, Hampshire, GU30 7JQ, England, (011) 44 (428) 722051, Est 1962

Grayshott Hall Health and Leisure Center: Grayshott, near Hindhead, Surrey GU26 6JJ England, (011) 44 (42873) 4331 Est. 1966

Jamaica

The Sans Souci Hotel Club and Spa (Charlie's Spa"): P.O. Box 103, Ocho Rios, Jamaica, West Indies, (800) 237-3237, Est. 1930

Italy

Abano Terme; Montegrotto Terme; Battaglia: Three separate spas, famous since the days of the Roman Empire offer what some consider the very best of natural hot mineral and mud spas. Arrange separate lodgings at one of these two hotels;
Hotel Bristol Buja, Via Monteortone 2,35031 Albano Terme, Padua / Padova, Italy: (011) 39 (49) 669 390
Hotel Sheraton Orologio, Viale delle Terme, 66, Abano Terme, Padua / Padova, Italy; (011) 39 (49) 669 502
The island of Ischia on the western side of the Bay of Naples; This magical island is the Disneyland of spas. Along with Greek and Roman temples Ischia's main spas are: The thermal centers at Porto d'Ischia; Forio d'Ischia; Casamicciola; the Olmitello spring; and Lacco Ameno. Arrange this trip with your travel agent - it's well worth all the effort.
Montecatini is a legendary spa town of 9 thermal springs and over 800 hotels, pensions and palaces. Located 24 miles from Florence: I recommend the Grand Hotel e La Pace , Via Della Torretta 1/A, 51016 Montecatini, Terme, Italy, (011) 39 (572) 75801

France

Centre de Thalasso-Esthetique de Pointe Biotherm: Boulevard de la Mer, 14800 Deauville, France (011) 33 (31) 98 48 11
Institut de Thalassotherapie Louison - Bobet / L 'Hotel Miramar: 11-13 rue Louisso-Bobet, Avenue de l'Imperatrice, Boite Postale 159, 64202 Biarritz Cedex, France (011) 33 (59) 24 20 80

Scotland

Stobo Castle Health Spa: Stobo Castle, Peebleshire, EH 45 8NY, Scotland (011) 44 (7216) 249 Est. 1978

GLOSSARY
Author's note:
Definitions apply to text of *Eliminating the Stress of Weight Loss.*

A

Acu-Point Facial Toning Points - specific nerve endings on the surface of the skin that cause a toning effect on underlying muscles.

Acu-Points - specific nerve endings located on the surface of the skin that create a specific effect when stimulated.

Acupressure - the application of pressure to specific nerve endings located on the surface of the skin.

Acupressurist - a therapist who performs acupressure treatment.

Addiction Erasure - the self-treatment needed to erase the craving impulse of addictive substances.

Adrenalin - a hormone secreted by the adrenal glands.

Affirmation - a positive declaration that something is already so.

Antibodies - substances produced by the body to counterbalance the effects of a negative, stress-producing agent.

Anxiety - fear and worry about a projected future event that may or may not happen.

Applied Kinesiology - the study of the relationship between stress producing agents/events and the muscle/nerve complex.

Art Therapist - a person trained in guiding an individual through a crisis using the therapeutic aspects of color and art.

Automatic Motor Response - a learned, conditioned action that involves the involuntary use of the muscular system.

Automatic Pilot Response - the circuit set up in the nervous system to facilitate the correct "conditioned response".

B

Behavioral Pattern - a recognizable consistent automatic reaction in response to a specific situation.

Biofeedback - a phenomena that exists within the nervous system that allows for feedback signals to regulate responses.

Biofeedback Reflex Test (B.R.T.) - a manual testing method used to monitor the Food Instinct responses to Food Rejectivity.

Biofeedback Response - the signal sent by the nervous system to regulate the response of a specific organ or system.

Bracing - see muscle bracing and muscular armor.

C

Calorie - a unit of measurement that represents the amount of heat "energy" provided by a specific food.

Central Nervous System - a unit consisting of the brain, spinal cord, and related branches that coordinates actions, thoughts and needs.

Chiropractic - a health profession that is concerned with the relationship between the nervous system and ill health.

Circulatory Response - one of the seven automatic responses to "stressors" in one's environment.

Cleansing Program, The - a 14-day program that clears the body of all toxins and waste debris generated by dieting.

Clinically Reproducible - obtaining the same results anywhere in the world by following a specific set of instructions.

Complex Motor Response - an automatic, conditioned, learned response that involves multiple organ systems.

Conditioned Programmed Behavior - pre-developed automatic actions taken by an individual without the need of directed thought.

Conditioned Response - the autonomic nervous system mechanisms needed to complete programmed behavior.

Conditioning - developing an automatic response to a situation by the prior repetitive completion of the desired response.

Cortisone - a hormone produced by the adrenal glands.

Cross-Crawling - a specific movement pattern designed to exercise the coordination of right and left brain patterns.

D

Defense Pathway Mechanism - the signal response activating those systems necessary to repulse the effects of a "stressor".

Dietary Distress - a negative response by the digestive mechanisms in response to a person's specific negative foods (see Stress Food).

Digestive Response - the way in which the body will prepare itself to process a specific food.

Digestion - a process involving digestion organs by which food is broken down mechanically and chemically.

Dysponetic Patterns - the involuntary use of the wrong conditioned reaction to a non-specific event.

E

Electromyography - the study and interpretation of a graphic record of a muscle's reaction to stimuli.

Emotional Memories - the stored, automatic responses of the nervous system to previous life events.

Empty Foods - foods that provide only a minimum or no vitamins, minerals or other essential nutrients.

Endocrine System Response - the response of glands that secrete substances into the body in response to stressful situations.

Endorphins - natural substances produced by the body that create a euphoric feeling in an individual.

Energy Reservoir - the maximum amount of energy available to an individual during any 24-hour period.

Energy Drain Syndrome - the physical and emotional complaints caused by a gradual depletion of the energy reservoir.

F

Facial Beauty - a conception of appearance that varies from one culture to the next; generally based on tone, complexion, color, etc.

Fat Cell Debris - waste products produced by the destruction of fat cells in the body.

Food Rejectivity Syndrome - the physical and emotional complaints caused by the ingestion of an Unsuspected Stress Food.

Food Stressors - negative foods that sabotage weight loss dieting.

Food Instinct - the ability (intuition) to sense which foods are OK to eat and which food are to be avoided.

Food Instinct Monitor - a biofeedback tool used to detect negative and non-foods during weight loss dieting.

Food Addiction - a neurologically conditioned response pathway mechanism set up by the body in reaction to a negative food.

Food Craving - the interpretation of the feelings produced when the response pathway of a specific negative food is triggered.

Foreign Element - those negative components of a food that trigger the Food Rejectivity Syndrome.

Four Environments - the chemical, emotional, external and structural spheres in which all humans must function.

Fuel - food consumed to generate energy.

Future Distress - present day physical/emotional problems generated by fear and worry about a future event that may or may not happen.

G

Genetic Blueprints (RNA) - the specific arrangement of the components of the RNA molecule that determines our individuality.

H

Healer Within, The - the autonomic nervous system which regulates the repair mechanisms of the body.

Health Enhancement - the process involved in the transition from a state of illness to one of better health.

Heredity - one's personal characteristics created from a blending of the character traits of both of one's parents.

Hidden Splendor - the uniqueness that represents the "spirit" within each person's soul.

Human Potential - the capability or capacity to change oneself.

Hypoglycemia - a state of discomfort produced by irregularities of a person's blood sugar levels.

I

Immune System - those mechanisms of the body that are responsible for enhancing one's resistance to negative factors in one's life.

In the Clear - the act of completing the Biofeedback Reflex Test with no food in one's mouth.

Incorrect Trigger - a stimulus that causes the body to respond with an incorrect action.

Industrial Diet - food trends and fads of the Industrial Society (the current commercial culture).

Inner Awareness - the ability to recognize the internal signaling patterns generated by the body in response to a stimulant.

Intuition - the capacity of knowing without the use of rational processes.

L

Leukocytosis - the increase in number of leukocytes (white blood corpuscles) in the blood generally caused by a foreign element.

Lymphocytes - special cells created by the body in response to foreign elements.

M

Mind Power - the conscious choice to direct the functions of the body.

Mind's Eye - the T.V. screen within the brain that converts words that are heard into pictures perceived by the brain.

Muscle Response - the reaction of the muscular system in reaction to various stimuli.

Muscle Bracing - an automatic response by the muscles of the body in response to stress-producing situations.

Muscular Armor - a state of preparation for action produced by the bracing of muscles.

N

Naturopath - a physician trained in the use of natural remedies to create a healthy condition for patients.

Negative Behavior - a behavior pattern generally rooted in prior conditioning that subconsciously sabotages present goals.

Neuro-Muscular Holding Point - specific nerve ending on the surface of the skin that, when stimulated, releases anxiety, stress and tension.

Neurology - the study of the nervous system as it relates to health and disease.

Neurologically Ingrained - physical internal pathways that store, for future use, conditioned responses to foods, thoughts or events.

Non-Foods - foods that when broken down by the digestive system, give up no chemical elements that can be used by the body.

Nutritionist - a person trained in the relationships between the components of food sources and human body needs.

P

Parasympathetic System - the system that controls involuntary actions such as breathing, heartbeat, repair of wounds, digestion, etc.

Physiological - concerning body (muscle and organ) function.

Physiological Response - the reaction of the physical body in response to various stimuli.

Pituitary Gland - a gland that secretes hormones that regulate many bodily processes, i.e., growth, reproduction.

Postprandial - events that occur after eating.

Power of the Subconscious - the ability of the autonomic nervous system to automatically control functions of the human body.

Protein Molecule - a component of all life-giving healthy foods.

Psychology - the science devoted to the study of mental processes, both normal and abnormal.

Psychological Response - an individual's perceived interpretations of and subsequent reactions to stimuli based on these perceptions.

Psychotherapy - treatment of dysfunction limited to interview and discussion techniques.

R

Refined Sugar Cycle - a series of reactions produced by the interreactions between the pancreas gland, insulin and blood sugars.

Rejectivity Defense - a series of reactions produced in the body in response to Food Rejectivity.

Resistive Capability - the ability of the body to mount a defense against negative life events.

Respiratory Response - the reactions of the breathing mechanisms in response to stress-producing stimuli.

Response Pathway Mechanism - the triggering of all necessary organs to insure successful completion of a reaction to stimuli.

S

Sabotage - an action that hinders or defeats an endeavor.

Scavenger Cells - cells from the immune system that are sent throughout the body to isolate and remove foreign elements.

Self Image - a perception based on prior conditioning that determines how one envisions himself or herself.

Self-Determination - the ability to make the conscious decision to change a specific aspect of one's life.

Self-Treatment - the ability to assume the reponsibility for one's own health by learning what is necessary to complete a goal.

Stress Food - a food, when eaten, that causes a triggering of the body's food rejectivity defense mechanisms.

Stressor - any agent or substance that triggers the Response Pathway Mechanism.

Sympathetic Nervous System - the nervous system that controls the functioning of involuntary muscles associated with organ function.

T

Toxic Substance - any element that triggers the body's Defense Pathway Mechanisms.

Toxic Food - an ingested substance that, when broken down in digestion, reveals elements that trigger the Defense Pathway Mechanisms.

Trigger - a stimulus that initiates other actions or events.

U

Unconscious Memory - neurological pathways, created by prior events, that are stored for future use in the nervous system.

Unsuspected Stress Foods - good foods that produce a negative biofeedback response because of lifestyle, stress, heredity or illness.

V

Vitamins - naturally occurring chemicals classified as amines that are vital for normal metabolism, growth and development.

W

Waste Debris - remnants of fat cells destroyed during weight loss.

INDEX

PROGRAMS, BOOKS & SERVICES FROM THE MONTEREY WELLNESS CENTER

For your health care library, The Monterey Wellness Center is proud to offer these seven programs. Each is designed to tackle a specific health problem by showing you how to achieve the personal power and control needed to eliminate that problem from your life. By using your own hands, heart and mind to create greater health, you will begin to recognize and honor your own personal power. It is this recognition of your personal power that will allow you to maintain your health, strength and vitality for the rest of your life. The gift of health comes from what you do for yourself - not what others do to you.

Four of these programs developed by Dr. Shain are also available as audio or video albums - complete with instructional booklets.

P.S.

We are so sure that you will love these programs that they come with a FULL ONE YEAR guarantee.

There is absolutely no risk because The Center maintains a FULL SATISFACTION GUARANTEED RETURN POLICY.

Within **One Year** after receipt of your order, you may request your money refunded (minus the shipping charges). To receive credit or a full refund, just include your invoice number (to confirm the purchased price) with the returned merchandise (in resalable condition). No hassle. Just a cheerful refund.

*We are all blessed with
a natural, built-in healing
mechanism-
"The Healer Within".*

*This healing potential
merely awaits the
hand, the heart and the mind -
recovery is a natural heritage.*

*These five programs were
developed by Dr. Joel S. Shain
to help readers
discover their natural heritage.*

How to Survive
the Poisoning of America
The Toxic Chemical Diet

America's food industry uses 8,627 different food additives. Foods are artificially dyed, preserved and flavored. They are injected with a wide range of growth hormones, ripeners, emulsifiers, texturizers, stabilizers and sweetners. And let's not forget the thousands of pesticides sprayed on our fruits and vegetables. This book teaches readers a fascinating biofeedback technique that the entire family can use. Finally, here is an effective method to discover if you are one of the 60 million people who suffer from a wide range of reactions to chemical additives. The book shows readers how to discover which of the foods in their diet are causing them physical distress.　120 pages; $9.95

Life Empowerment
*Eliminating the Problem of
The Chronic Energy Drain Syndrome*

Most people are unaware of the great wealth of energy, vitality and strength stored within their bodies. They live 90% of their lives suffering from Chronic Energy Drain Syndrome - experiencing rare, occasional moments of true vitality and energy. One of the more popular classes at The Wellness Center, this book presents a compilation of the most successful ancient and modern techniques that will give you the necessary keys to tap into your vast reservoir of natural vitality, strength and energy. Every Dis-Ease known to man, starts as a breakdown in this reservoir. Restoring optimum joy, vigor, life and health must truly start by restoring these factors. This manual shows readers how to unlock their bodies' revervoir of optimum health. Used by Olympic athletes at the 1986, 1988 and 1992 Olympics. 230 pages; $14.95

The Whiplash Syndrome
Erasing The Trauma Whiplash

A whiplash injury is one of the least understood of all traumatic injuries. Medical studies have shown that 60% of all auto accident victims suffer complaints for many years: some for the rest of his or her life. Yet, this trauma can be corrected easily if the victim is given the correct tools. This book is for the whiplash victim who continues to suffer distress from his injury. It contains a comprehensive self-treatment manual designed to teach the techniques needed to soothe and heal the problems that result from a whiplash injury. *The Whiplash Syndrome* includes an insightful explanation of why the victim suffers and why self-care is the fastest, most effective and most long-lasting of any treatments. 230 pages; $16.95

Life Without Arthritis
The Arthritis Self-Care Program of
The Monterey Wellness Center

Arthritis is a degenerative disease. If nothing is done to directly soothe, heal and restore the troubled arthritic area, it continues to degenerate. Pain drugs only sedate your awareness of the problem - they do nothing to correct the problem. *Life Without Arthritis* offers the arthritis sufferer a safe and sane alternative to drugs. You can choose to come to the aid of your arthritic body. This highly successful program from The Monterey Wellness Center teaches marvelous self-care remedies for all forms of osteoarthritis. Arthritis sufferers can prevent the problem from worsening (further degeneration) while erasing pain and restoring function. 230 pages; $16.95

Eliminating
The Stress of Weight Loss
Recognition and Control of
Stress-Induced Weight-Loss Saboteurs

This fascinating health enhancement program from the Monterey Wellness Center provides essential information for anyone wishing to control and/or maintain hard-earned weight loss achievements. It is designed to maximize the potential of all current weight loss regimens by confronting life's daily struggles such as anxiety, stress, tension and prior conditioning. Weight loss goals are destroyed when dieter's cannot control their body's Natural Stress Response Mechanism as it reacts to these daily struggles. This book shows dieters how to recognize and then eliminate their personal stress-induced weight loss saboteurs. These exciting new ideas and concepts show the dieter how to avoid the anti-weight-loss effects of daily stress. 232 pages; $14.95

Smart'n Up
A Self-Improvement Program
for the Mind by Dr. Barbara B. Brown

Scientists say, "We use only 10% of our brains." Do you know what that means? - that we have some 90% more brain power to use if we only knew how! Imagine what you could do if you could use 25% of your brain - or 50% or 75%!

Dr. Brown has authored *New Mind, New Body* (1974); *Stress and the Art of Biofeedback* (1976); *Supermind, the Ultimate Energy* (1980); and *Between Health and Illness* (1984).

This new book from Dr. Brown is the first program ever developed to enhance intelligence levels. It accomplishes this by providing 20 imaginative mind exercises designed to access the genius lock within this 90% of the brain. By describing the operation of the subliminal intellect, Dr. Brown is able to demonstrate to the reader the ease with which super intelligence can be accomplished. 266 pages; $14.95

Odyssey of Awakening
A Poetic Journey by Susan Elizabeth -

This book provides a roadmap to wholeness in flowing poetic form. A deep exploration of what it means to be human, this compelling and powerful work takes us from the agony of abuse to the ecstasy of inspiration. This is done with such loving care that we come out of the experience vastly enriched and armed with new tools with which to overcome our own inner demons. Overflowing with insight, the words linger on and blossom with ever-deeper meaning that expands with each reading. So powerful are the concepts that we are infused with strength and courage and bathed in compassion, leaving us refreshed and sated every time. Filled with endless "aha's", this triumphant book is truly a guiding light. 143 pages; $12.95

222

TWO BIOFEEDBACK MONITORS TO HELP DEVELOP
YOUR FOOD INSTINCT SIGNAL

The Jamar® Dynamometer (pages 15, 68, 175 and 177) is the perfect tool for those individuals who wish to monitor their Food Instinct Signal by themselves - without the aid of a friend. It is a precision medical instrument designed for evaluation of nerve/muscle injury. With this tool that measures grip strength from 0 to 200 pounds, the dieter can develop their Food Instinct Signal by observing fluctuations in their grip strength caused by "Unsuspected" Fat-Producing Stress Foods. Anyone may call their local medical supply dealer to purchase a monitor.

The Food Instinct Monitor (pages 70 and 180) offers dieters an inexpensive self-monitoring biofeedback tool to identify stress foods. Similar to the Jamar Dynamometer in concept, it is made of heavy duty durable molded plastic. This monitor is small enough to slip into your pocket and take anywhere. Available only from The Center for $25.00

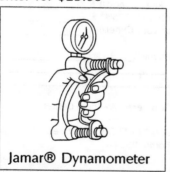

Jamar® Dynamometer

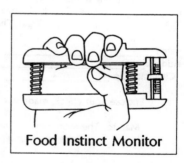

Food Instinct Monitor

A Free Gift With All Orders
The Biofeedback Reflex Test Instructional Booklet

As our way of saying, "Thank you for your order.", we will include FREE with all orders, "*The Biofeedback Reflex Test Instructional Booklet*". This value-packed booklet provides all of the directions and pictures needed to learn this fascinating procedure. This booklet is designed to be easily reproduced on all copiers. Please feel free to share this with your friends, relatives and co-workers.

SO THAT OTHERS WILL BE ABLE TO ORDER THIS BOOK,
PLEASE PHOTOCOPY THIS PAGE BEFORE PLACING YOUR ORDER
OR
SEND THE NECESSARY INFORMATION ON PLAIN PAPER
ALONG WITH YOUR CHECK OR MONEY ORDER.

Complete One Year Money-Back Guarantee!

	Cost	Quantity	Total
Eliminating The Stress of Weight Loss	$14.95		

Shipping & Handling: ($3 first item: $2 each add'l) $_____
Sales Tax: (California residents add 7.75%) $_____
Total (Thanks For Your Order!) $_____

Name _____ Phone (____)_____

Address_____

City/State/Zip _____

Enclose your check, made out to The Monterey Wellness Center

Mail to: The Monterey Wellness Center, 335 El Dorado Street,
Suite #8, Monterey, California 93940.

IF PAYING BY CREDIT CARD:
Master Card / Visa # _____

Expiration Date _____ Signature _____

For quantity discount prices or rush UPS handling,

Call Toll Free- (800) 347-1527

THIS BOOK IS AVAILABLE AT SPECIAL QUANTITY DISCOUNTS FOR BULK
PURCHASES FOR SALES PROMOTIONS, PREMIUMS OR FUND RAISING.

FOR DETAILS WRITE TO:

THE MONTEREY WELLNESS CENTER

335 EL DORADO ST., MONTEREY, CA. 93940

OR CALL:

1 (800) 347-1527

Dr. Joel S. Shain

Dr. Joel S. Shain is an eminent physician, educator and lecturer who is considered one of the country's leading experts in the field of health and wellness. Since 1967, Dr. Shain has written and created health programs that have literally shaped the thinking of generations of Americans. He is responsible for creating the science and health programs that are taught in many elemantary schools in New Jersey as well as many instructional programs that are taught to physicians throughout the country.

In 1978, Dr. Shain created The Monterey Wellness Center in Monterey, California. This health care facility was staffed by Medical Doctors, Chiropractic Physicians, Naturopathic Physicians, Acupuncturists, art therapists, biofeedback experts and guest lecturers/healers who came to the center to share their knowledge. To help patients in their personal transformation from illness to vibrant good health, Dr. Shain researched, tested and proved effective combinations of safe and effective home health improvement techniques that were taught at the Center. Not only did these programs help in the patient's transformation to vibrant good health -- the programs were also the means by which the patients stayed well after leaving the Center. This book is one of the many programs researched, developed and taught by Dr. Shain at the Center. In each of his books and programs, he discusses a specific health problem, and then presents home self-care techniques needed to soothe, heal and restore the body.

Dr. Joel S. Shain is the author of:
Eliminating the Stress of Weight Loss
How to Survive the Poisoning of America
The Whiplash Syndrome - Erasing the Trauma of Whiplash
The Pain Erasure Program of the Monterey Wellness Center
Life Empowerment - Eliminating the Chronic Energy Drain Syndrome
Life Without Arthritis - How to Soothe, Heal and Restore the Arthritic Body

Dr. Shain has also written, coordinated and instructed health and healing programs for the following organizations: The Touch For Health Foundation; The Dental Kinesiology Foundation; The Sacro Occipital Teacher's Organization; and The National Center For The Transitional Healing Arts.

Fascinating New Concepts
Controversial New Ideas
Exciting & Stimulating

Information That Will Allow You To;

Easily Achieve and Maintain Weight Loss Goals

Eliminate Dieting's Stressful Toxic Waste Back-Up Problems

Discover and Eliminate Unsuspected Fat Producing Stress Foods

Eliminate Food Addictions Caused by Anxiety, Stress and Tension

Eliminate Facial Lines and Wrinkles Caused by Stress and Diet

Eliminate Negative Eating Reactions Caused by Anxiety, Stress and Tension